Intelligence in Energy

To our beloved families

Intelligence in Energy

Gülgün Kayakutlu
Eunika Mercier-Laurent

iSTE PRESS

ELSEVIER

First published 2017 in Great Britain and the United States by ISTE Press Ltd and Elsevier Ltd

ISTE Press Ltd
27-37 St George's Road
London SW19 4EU
UK

www.iste.co.uk

Elsevier Ltd
The Boulevard, Langford Lane
Kidlington, Oxford, OX5 1GB
UK

www.elsevier.com

British Library Cataloguing-in-Publication Data
A CIP record for this book is available from the British Library
Library of Congress Cataloging in Publication Data
A catalog record for this book is available from the Library of Congress
ISBN 978-1-78548-039-3

Printed and bound in the UK and US

Contents

Acknowledgments

We would like to present our gratitude to Dr. Debra Amidon, founder of ENTOVATION Network, for having connected us.

Foreword

Intelligence in energy is a very important issue for energy management, production and consumption. To face climate change, the attitudes of all involved actors have to evolve.

This book presents the above topic in the large spectrum – from history of heating, through energy status in developed and developing countries to different optimization methods urgently required for today and future energy dynamics preserving our living conditions.

The modern world is a huge consumer of energy. Newly discovered materials allow producing devices with significantly lower consumption per unit of energy. However, technologies are entering into newer areas of life. New countries join the developed and developing ones what increase the global demand for energy. The current time parameters (daily, annual) of energy consumption are changing. The exponential development of power electronics increases the demand for reactive power and raises new critical situations in power systems.

Conventional power plants influence the climatic change (warming), nuclear power plants afraid and have many enemies. Alternative energy sources are only a supplement to the classic ones. Renewable energy sources (RES) are not providing electricity continuously – wind is not always blowing, resources of water in reservoirs are not always sufficient; the Sun is not shining in the night. These resources are not able to ensure the continuity of supply despite the development of methods for energy storage.

Small systems are able to provide a flexibility of energy supply by acting independently. Large systems must ensure continuity despite the unpredictability of renewable energy sources which needs to be held in readiness of entire blocks of energy. In this changing context intelligent optimization techniques such as artificial and computational intelligence may play a significant role.

This book discusses the possibilities of energy generation ad use by the wide range of companies and organizations today and provides some perspective for the future. Main methods of optimization are presented as well as the main intelligent techniques having the potential to improve the life cycle of energy. The focus on training and education is extremely important to integrate new resources and technologies, but also to influence behaviours of intelligent use of energy through serious games.

Excellent perspectives on the future of the exponentially growing cities is also presented and propose some thoughts encouraging the readers involved in the cities to look for alternative solutions.

The implementing of Knowledge Management methods to optimize the use of energy sources and the demand for energy represents a great added value to current energy management systems. It will allow the real optimization and dynamic management in function of needs that are not easy to plan today.

Taking into account the different degree of economic development of all countries, available and prospective resources, degree of unfolding transmission networks, the possibility of exchange - the covered topic is very broad.

I wish all readers of this book to gain another view of energy life cycle, to think beyond the obvious, and realize that we can do better to protect our planet.

MAREK SUPIK

November 2016

Introduction

"We must leave oil before it leaves us"

Fatih Birol, Head of International Energy Agency

Energy has always been associated with the universe and the latter with living things and humans. Physical sciences have indentified many forms of energy such as kinetic, chemical, magnetic, gravitational, elastic and radiant energy that collaborate with one another.

As we did not yet understand how to wisely use the energy from the universe, we discovered natural phenomena and resources to provide energy for powering increasingly numerous devices that we invented. As the natural balance of the biosphere was not respected, we now have to fix the damages that we created. The concept of renewable energy is not new; however, considering the current challenges, it is brought up-to-date.

The motivation for writing this book came from daily exchanges with energy experts in the industry and with our students. During a visit to a manufacturing company, an energy manager said "I do not need any load management software for scheduling the four natural gas tribunes that we are running; they work with full capacity 7 days 24 hours". It is obvious that an accountant has not yet shown the cost comparison to the managers. In other words, the investor has not yet seen that the full capacity usage means more expense than full capacity production. Nor has he met any ecologist inspector who taxes inefficient use of energy.

Considering the fact that energy is essential for today's society, we wish to share with younger generations our thoughts and experiences and empower them with the ability to wisely manage scarcity and think about impacts when reviewing improvements on the subject.

One of the greatest changes in human history has been the unprecedented, rapid increase in population, due, among other reasons, to medical progress. We are now able to face serious diseases that we created. The worse human health is, the higher the energy demand. Global forecasts show that the energy demand will reach 1 zetajoule in 50 years.

The rapid increase in energy demand, especially in developing countries such as China and India, is a major concern. Dependence on other countries due to energy importation has always been one of the enemies for national economies. Even after using shale gas, the USA has more than 40% importation, as reported by the National Research Council. Both fossil fuel owners and importers are worried about the reliability on traditional sources, besides the issue of climate change caused by combustion. The volatility of energy prices has been causing a bullwhip effect on energy investments since the 1970s.

All of these issues have slowed down the investments in energy infrastructures and related technologies. Alternative energy sources have not yet been considered. Today, a few square kilometers of forests per year are still being cut for heating; more than 50% of energy is contributed by coal. Energy generation is inefficient, disconnected and centralized.

Nuclear power is considered a clean energy source, but its associated risks have to be evaluated. Recently, Ukraine and Japan were the victims of such risks. The causes of the Chernobyl disaster are different from those of Fukushima. The first is due to the poor means allocated for maintenance, while the second one is a result of linear thinking and money saving. Except for the ambitious magnetic fusion project ITER, this is one of the rare industries where the investor's vision has changed very little since the coal mining days. In 2014, more than 300 miners died in the Soma coal mine accident in Turkey because coal investors wanted the full capacity work of manpower without enough security. Cost optimization in this industry is made only with respect to employee reduction.

The ubiquity of technology use often prevents energy users from taking effective action when rising prices provide a strong economic motivation to save energy. This is mainly because the history of material progress over the last century has made energy sources less costly in real terms and has made energy flows invisible to energy users. This has led to two major consequences, namely energy unawareness and energy invisibility.

This book presents a knowledge-based review of past, present and future energy uses. Energy is seen here from various perspectives: mathematical, energy process modeling, economic, technological and environmental.

Chapter 1 presents a historical sketch of energy in human life. It puts forward a short history of heating and harnessing natural resources by humans. It then presents the current landscape of energy in the world from various perspectives. It also discusses the influence of technological progress in energy lifecycles and ends by setting the scene for environmental constraints that are the enablers of innovation.

Chapter 2 provides an overview of current trends, components and technologies for energy generation and distribution. It then reviews the main optimization techniques. Finally, it covers uncertainty handling.

Chapter 3 describes a large spectrum of "intelligent" technologies useful for energy lifecycles as well as in other related fields. Major artificial intelligence techniques are described here and some examples of applications are given. Finally, it highlights the art of combining various techniques for solving complex problems effectively.

Chapter 4 provides a review of international greening commitments with their impacts on macro-energy balancing models. It also underlines the role of training and innovation in converting commitments into action.

Finally, Chapter 5 presents some current experiments, trends and projects that open the door for a new way of thinking in energy use.

We hope that this book will allow readers to consider energy differently and will inspire the evolution of existing professions and the creation of new ones in the energy field.

We would like to promote the wise use of energy and intelligent techniques to manage it, an alternative way of considering energy from various perspectives and as a system component of our natural ecosystems.

We expect your feedback on http://intelligence-in-energy.eu

1

Humans and Energy

1.1. History of heating

In ancient times, the Sun was considered as God. In Mesopotamia, in 4500 BCE, houses were built from sun-dried bricks; the exterior was "white washed" to serve as a defense to radiant heat. Small lamps fueled by sesame oil provided lighting. After the month of September in rainy season, houses were heated by burning palm fronds or palm wood.

Korean and Roman societies were using under-floor radiant heating thousands of years ago. The Romans used a system of hypocausts, which pumped heated air into open spaces beneath raised buildings. Koreans used a radiant heating called "ondol" as far back as 1,000 BC, where a wood-burning furnace at one end of the home pumped hot air through the channels under the home, to a chimney on the other side (radiant heating is rumored to be the reason why Koreans take off their shoes when entering a home and sit on the floor) [GUE 13].

Humans have always explored the available resources, sometimes adapting them to the local climate.

European homes in the Medieval Ages were still heated primarily with fire on open hearths. Medieval homebuilders used a clay and straw (or "wattle and daub") home insulation technique that was both efficient and environmental-friendly; in fact, it is a comeback among today's green constructors.

In the 14th Century, European homes and castles adopted the use of chimneys – a dedicated hole in the ceiling – which allowed hearths to be moved from the center of the room to places against the wall. This led to a significant improvement in living spaces, but the popularity of chimneys spread slowly; they were still relatively uncommon 200 years later.

The next major stride in home heating came with the invention of the stove in the late 17th Century. Stoves became increasingly popular in the next two centuries, where wood or coal was used as the primary fuel. Coal, in particular, generated an especially large amount of soot that eventually proved to be detrimental to the environment; sulfur dioxide, a by-product of burning coal, is largely responsible for acid rain, and coal-burning stoves regularly left black stains on walls and ceilings. COP21, which took place in Paris in 2015, put emphasis on reducing coal and coke coal heating as well as on replacing coal power plants by renewable energies.

The ceramic stove was probably invented in the 18th Century with the objective of consuming less wood or coal. With the first energy crisis, these stoves became very popular in cold countries.

Many houses are still using radiators invented in the 19th Century. However, the recent trend is to go independent, combining locally available renewable energies as discussed later.

1.2. Landscape of energy

The global energy landscape is undergoing radical changes. The limits of fossil energy resources are expanded with technological improvements and research focused on renewable energies. Shale gas discoveries in the USA and Mexican privatization of oil delivery have led to an increase in energy supply and caused an unexpected drop in prices. This has become the joy of North American countries, which are moving closer to independence in energy supply.

European countries import energy from North African countries and Russia. Yet, Europe has made a great improvement in energy use so that energy consumption in 2013 was exactly the same as in 1990 despite economic growth. German success with renewable energies has opened the door for removing nuclear energy plants. France has improved smart cities using renewable energies, although 18 nuclear plants are still kept. Furthermore, energy security in Europe is improved.

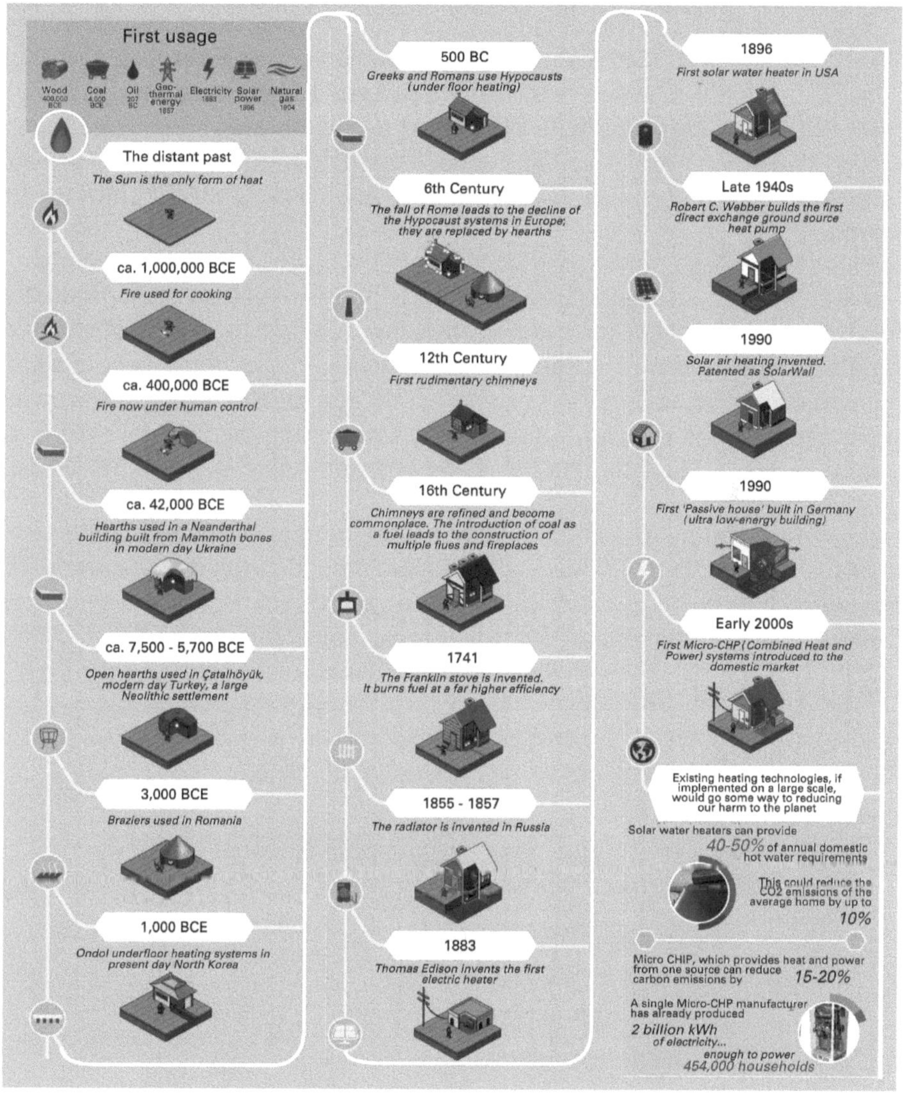

Figure 1.1. *A history of heating. Modified from [HH 13]*

Energy specialists have identified China, India, South Korea, Singapore and the rest of the developing Asian countries as the largest growing energy consumers of the globe. Until 2040, a 60% growth is expected in that region. The USA and China are collaborating on the biggest resources of shale gas in China.

With respect to price reduction, the OPEC is expected to cut the energy supply. No cut has been seen in Saudi Arabia or Libya even during their political instabilities. China has decided to take a bigger role in the Middle East. China's energy ties with Iraq, Iran and even Kuwait emerged from strong foundations of service and labor exports. Both the Middle East and Latin American oil producing countries are paying back the Chinese credits in crude oil.

Fatih Birol, the head of International Energy Agency, states that "As South East Asian countries develop, they are moving towards the center of the global energy state". Although individual energy consumption rates are below the average, energy is consumed in the region and demand is growing in parallel with the increasing power use. Connecting the international grids will be an option to respond to the needs of South Asian countries [HAR 16].

Discussion on hydrocarbon potential is ongoing between South American countries such as Brazil, Venezuela, Argentina and Colombia. Explorations may allow the region to have a more important role in the energy landscape. However, security issues and the technological challenges continue.

The Australian government has recently established an aggressive target to increase efficiency in energy production as a self-sufficient energy consumer.

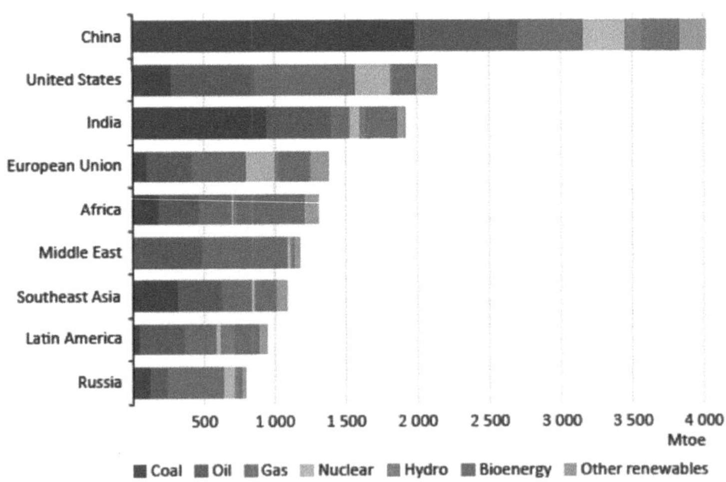

Figure 1.2. *Prime Energy demand in 2040*

India is the fourth largest energy consumer in the world and its GDP growth is observed to depend on how well the country manages its energy industry. A series of political reforms announced in 2014 have attracted investors in oil and gas industries. New revenue sharing models have been developed for renewable investments, since rooftop panels are seen as the best solution to avoid service issues.

South Africa is converting coal plants into renewable energy investments. The rest of the African countries are hoping to overcome scarcity through the newly discovered energy industry. Ghana has just discovered oil resources and is planning to double its production every year. Dropping oil prices and the decrease in Chinese development has made 2016 tough for Africa. More than 700 million people in Saharan Africa still lack electricity [INT 14c].

Energy users should observe these changes in the energy landscape as a greater opportunity to respond to the needs of the growing population and to avoid the volatility of prices while creating new jobs. This can be seen as a new mission of supplying the energy that powers modern life. More energy supply with reduced prices will also be a tool to reduce the economic gap in the globe.

1.2.1. *Advances in North America*

Advances in equipment and drilling in deeper offshore waters and soil have increased oil reserves in the USA. Ocean vessels have contributed to the efficiency of delivering liquefied natural gas. However, the main success is brought by hydraulic fracturing and horizontal drilling to extract oil and natural gas from shale rock. Furthermore, oil sands in Canada have been made more beneficial through the use of steam-assisted gravity drainage and other techniques [ERN 13]. As a result of those advances, the IEA expects the USA to improve oil and gas reserves by almost 20% until 2035. With the new shale gas and oil sand resources, North America is expected to become an independent oil supplier by 2025. The addition of private sector know-how has also increased production by 25% or more in a decade. This would contribute to a rapid increase in oil and gas production in the USA and Canada. Furthermore, experiences in shale gas operations will be used by other shale gas producers like China.

Hence, the USA has increased the domestic production and supply of oil and gas, which has reduced market prices and improved the American economy. Unexpectedly, a large amount of solar panel requests from China has also contributed to Californian producers. Incentives have enabled the wind power producers to use computer-aided design, computer-aided manufacturing, planning, simulation and analysis tools in order to minimize the production costs. Analytic tools are used to follow the trends close enough to avoid vast changes that would harm the market balance.

Lithium supply is no longer a risk for power storage and electric cars in North America, because Australia and Chile are no longer alone, Zimbabwe and Argentine have also taken roles in lithium production. Moreover, Tesla is investing in production of lower cost batteries.

The fact that Mexico allows private investment in energy production is also very important for North America. Mexican oil is expected to be 3 million bpd including the deep offshore resources. Mexican solar and wind resources will also be managed by American and Canadian companies.

1.2.2. *Implementation of new energy policies in Europe*

The renewable energy revolution and the considerable decline in energy demand due to efficient use are changing the energy landscape in Europe. The supply uncertainty caused by Russia and North African countries continues. Russia, Norway, Saudi Arabia, Nigeria, Libya, Algeria and Qatar are the major energy exporters to the EU, which adds more difficulty to the issues of creating collaborative policies.

The EU Commission for Energy and Environment overviewed the collaborative policies in July 2016. A Member of the European Parliament pointed out the emergence of a new legislative package to empower the renewable energies [EUR 16a]. It is observed that the real European strength in energy policies is based on the cities. On the one hand, consumers having the energy-efficiency culture should be given new financial tools to take individual actions. On the other hand, smart grids supported by local managements will spread over to large landscapes. Mary Donnelly, the European Commission's Director for Renewables, Research & Innovation, declared a smart financing tool for smart buildings, to support better risk assessments and energy-efficiency projects.

A more quantitative assessment of industrial sector externalities caused by collaboration would be valuable in informing priority setting for future initiatives to reduce the costs associated with industrial operations.

Further training will be supported to enhance green training, eco-designs and efficiency culture.

1.2.3. *New investments in China*

China is considered as the biggest energy supplier and consumer of our century, the primary resources of which are still coal and hydro energy. There is a trend started for conversion into clean energy sources including nuclear energy.

All developed countries are further investing in new opportunities in China. For example, Australians are opening lithium production sites in China.

The USA and China are collaborating on encouraging higher efficiency and renewable technologies. Because of high CO_2 emission caused by coal use, the two countries are primarily focused on pollution. Following climate meetings, the USA and China are activating the potential of energy-efficiency improvements and enhancing efficient transportation systems. Improvements in buildings are being accelerated.

The governmental action plan in China for 2014–2020 aims to reduce coal consumption by 62% and increase natural by gas at least 10%. In the 2015 Foreign Investment Catalogue, the construction and operation of renewable energy power stations are encouraged. State loans are given to renewable energy suppliers for the local grids. In particular, hydro-power, solar panels and offshore wind farms are supported. A nuclear project will be started in 2020 [CHO 16].

Regulatory environment in China is changing the domestic market, opening the door for international collaborations to initiate the innovations.

China is enlarging its role in the Middle East, especially in the oil extraction technology that started with contradictions between the Middle East countries and the USA.

1.2.4. *In search of energy in India*

IEA provided detailed research on India and observed a transition towards an open system [INT 12] that defines the energy sector in India as follows: "The goal of providing energy access to the entire population led to well-meaning policies designed to protect the poor, but resulted in a system of untargeted producer and consumer subsidies that prevent a more thorough implementation of a well-functioning and financially sound energy sector". Currently, India's energy sector is unable to respond to the rapidly increasing power demand.

With the "Power for All" project, India expects to provide 24/7 electricity to all citizens by 2019 after having installed 10 GW of solar energy [PRA 15a]. Non-tax incentive policies and support for foreign investment in the renewable corridor broke the coal monopoly. Individual rooftops contain single energy plants with domestic solar panels. To provide access to energy in rural regions, mini-grid applications are widely used in rural areas [PAL 13], as shown in Figure 1.3.

Figure 1.3. *Example of a mini-grid [PAL 13]*

Well-known mini-grid samples can provide an average capacity of about 1 MWp, supplying stable and reliable electricity to about 10,000 households. Another example has electrified about 35,000 households across more than 1,400 villages and hamlets with low-capacity (1 – 6 kWp) solar mini-grids. Moreover, biomass gas using mini-grids are used in about 80 villages across different states in India.

Importation of technology and engineering services from Australia is expected soon. India's renewable power capacity is set to reach 170 GW by 2022. KPMG predicts that the cost of power generation by using solar energy will be 10 % cheaper than that using coal in 2020 [HOR 15]. The reduction of pollution and increases in GDP due to renewable energy investments are admirable in India.

The radical change in the energy landscape in India will be achieved once the smart grids and more planning are encouraged by the governmental policies.

Meanwhile, Indians are very wise in using their smartphones, especially in rural zones. Their producers adapt by providing long-term battery; for example Huawei sold in India 3,000 mAh batteries with saving mode. Rural zones are available for energy storage using the available batteries.

1.2.5. *Energy for Africa*

The context of wars and the hunger in Africa does not give way for energy investments. The number of people who have no access to electricity is an issue that should be handled by the international collaboration of all countries. IEA [INT 14b] predicts that 1 billion people will have access to electricity in Africa by 2040. However, because of the rapid population growth, 75% of the population will have no access to electricity.

Demand for energy continues to grow with increasing incomes in sub-Saharan Africa. Solid biomass is one of the major resources to respond to the demand. Clean cooking facilities using biomass will reduce residential air pollution. The majority of the people using these clean tools live in urban areas.

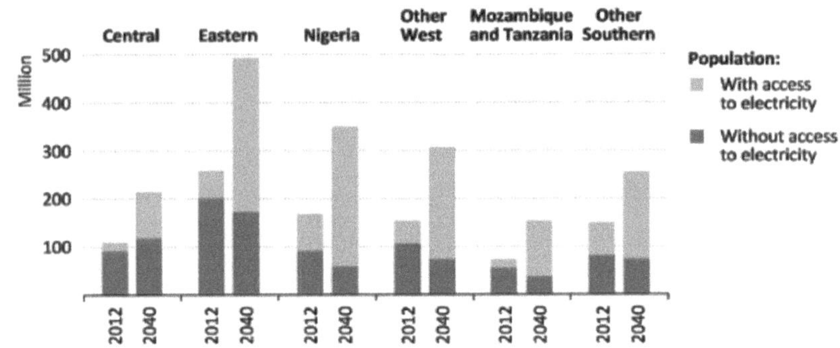

Figure 1.4. *Electricity use in Saharan Africa [INT 14c]*

The World Bank predicts that 50 billion USD of investment will be needed to respond to the power needs of sub-Saharan Africa, where the population is expected to reach 500 million in 2030 [GUA 14]. The European Commission was an early supporter of the UN project of Sustainable Energy for All and has been investing in Africa since 2002. China declared a loan of 20 billion USD in 2013, mainly for the infrastructure of the energy industry. The same year, China and the USA gave their support by providing 7 billion USD to install 10 GW of energy supply. As the second biggest oil importer in the world, China's investments in Africa are interpreted as creating independence from the Middle East oil [BRO 13].

A strategic analysis from Australia [PIE 16] indicates that drought threatens food, water and energy security because the lowest rainfall has been recorded since 1904. African people who are having difficulties in finding food would like to have food, water, light and heating at the same price: free of charge.

The vision of energy varies among the cultures. Some look at energy as a commodity, whereas others see it as an ecological resource, or even a strategic resource for society. Policy choices are implicit choices among different views. The selected policies are legitimized to have a cultural effect. When a society accepts a particular definition of energy, the terms for future political policies are set. However, energy is a global resource, which should be in the service of humans. There is a power game played on it between the business and the governments. Cities are trying to liberalize this resource.

1.3. Evolution of energy use in business

The industrial use of energy probably began with steam in 1900. That is where the waste of energy started as well. Ever since the pioneering research of Taylor, the relationship between technical progress and productivity growth has been envisaged within the interactions of cost and production. The cost of resources, including energy, has been one of the major concerns in several industries.

Since 1881, the coal-powered steam engine has been used to generate electricity for industrial use, just after Edison's discovery [SMI 94]. The energy use in industrial countries and farming countries has also widely differed [BUR 11].

Factory production halls were heated by heaters powered using gas blasts and later by water heaters. External combustion engines were next added to the water heater.

After the oil crisis in 1973, energy saving became the major cost reduction tool across the globe when energy steering committees were started in each nation. Industrial energy consumption took the biggest slice, i.e. more than 30% of the total energy use [INT 14a]. Table 1.1 provides the details of the changes in different prime energy uses between 1973 and 2012.

	Coal(b)	Crude Oil	Oil Products	Natural Gas	Biofuels and Waste	Other(c)	Total
1973 INDUSTRY	182.79	0.39	312.91	250.4	42.26	169.41	958.18
1973 TOTAL FINAL CONSUMPTION(a)	303.27	0.39	1,583.63	498.48	84.3	345.49	2,815.56
2012 INDUSTRY	95.51	2.33	96.69	250.92	66.98	280.49	792.92
2012 TOTAL FINAL CONSUMPTION(a)	120.66	7.07	1,697.43	709.15	180.77	867.41	3,582.49

Table 1.1. *Industrial use of different energy resources [BUR 11]*

a) Total Final consumption is the sum of Industrial use, and Transportation, Residential, commercial and Public use and Non-Energy Use

b) Coal includes shale oil

c) All other energy resources

The share of industry in a total of industrial, transport, residential, commercial, public and non-energy final consumption dropped from 34.01 to 22.13%. Industrial energy consumption takes into account iron and steel industry, chemical and petrochemical industry, but excludes feedstock, non-ferrous metals, non-metallic minerals, transport equipment, machinery covers, mining, food and tobacco, paper, pulp and printing, wood and wood products, construction, textile and leather, non-specified. It is surprising that even in 2012, hydro energy was not in industrial use, whereas, coal consumption increased from 60 to 79%. It is also interesting to see that electricity consumption dropped almost by 10% mainly because of the saving and efficiency investments.

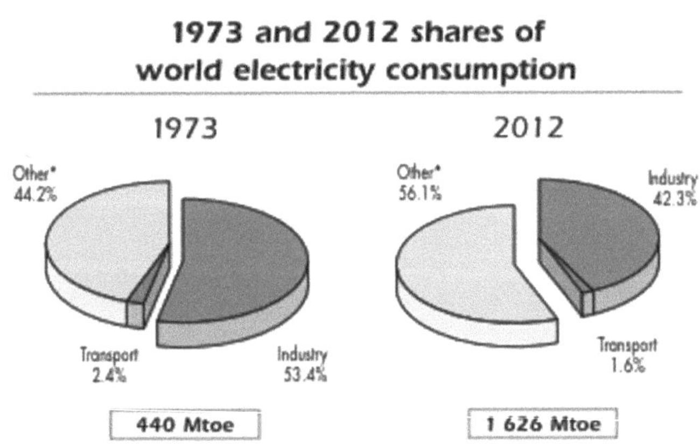

Figure 1.5. *Share of industry in power consumption [INT 14a]*

Labour and multifactor productivity concerns continued in the USA until energy prices began to fall in 1983 [BER 90]. Industrial energy use has had an important share in fossil energy use since the beginning with steam systems. The heating power has been the inevitable part of chemical processes.

The head of the Dutch Energy Steering Committee and a professor of physical chemistry, van Gool, was assigned the task of reducing the energy use in Holland. He focused on the energy optimization of industrial processes and applied the laws of thermodynamics to measure energy use [KUM 78]. In fact, to calculate the amount of energy needed to produce a certain amount of product in a heating process, energy conservation balance, the first rule of thermodynamics, is used. Hence, the sum of all the outputs from this system multiplied by their individual energy requirements must be equal to the sum of inputs multiplied by their individual requirements. This of course does not take into account the heating quality of the resource used. Such a mistake can be avoided by applying the second rule of thermodynamics by stating that not all the heat generated can be useful. It becomes necessary to define the parameters used, and exergy is a concept that came out of these processes as the available energy received from a resource under a certain environmental condition [HAM 07]. It is observed that the laws of thermodynamics can be used in any heating process measure in the industry.

World Outlook 2014 predicts that the industrial energy use will have the biggest growth (1.4% per year) until 2040, when compared with buildings, transportation and energy generation [INT 14b].

Foundation industries such as metal production, cement, lumber, and paper and pulp production have been the biggest consumers of energy. Energy consumption is based on the use of the kind of product, scale of production, structure of the input mix, technology and thermal efficiency of manufacturing machines.

It is not easy to design the energy system in a process plant, as it requires an in-depth understanding of each subsystem, as well as interactions of all sub-systems. To achieve an effective design, the physical insights of the processes are combined with the mathematical models of optimization. Methods have been developed for efficient operations and debottlenecking [ZHU 00].

Industrial revolution can be divided into four phases, starting from steam machines: mechanical revolution, electric revolution, electronic revolution and intelligent revolution [EUR 16b]. Energy demand has been observed to increase exponentially in the third phase where electronics is in the service of industry with micro-chips. However, none of the technologists of the 20th Century has predicted the effect of knowledge and intelligence uses that began in the 1990s.

Patrice Noailles defines the measures of the value of industrial innovations in three components: technical benefits, time value of innovation and shared value of innovation [PAT 11]:

– *Technical benefits* are achieved through the technology application to make production processes more effective and allow the production of more effective products. This is a value created for both the company and the society. This is because the company reduces the costs and uses less resource and the users of the product achieve more efficient results.

– *Time value* is the improvements made in time with incremental changes resulting from the feedback of a variety of users. There are innovations that multiply economic value by a thousand or a million in 20 years.

– *Shared value* is the increase in the number of beneficiaries of a single innovation. The new product/process is spread more widely if the beneficiary is the one who pays for it. On the web, the shared value has no economic definition as collective use is free of charge, but feedback shows that users have reduced satisfaction with the content.

The correlation between innovation and the energy use is already proved [GER 16].

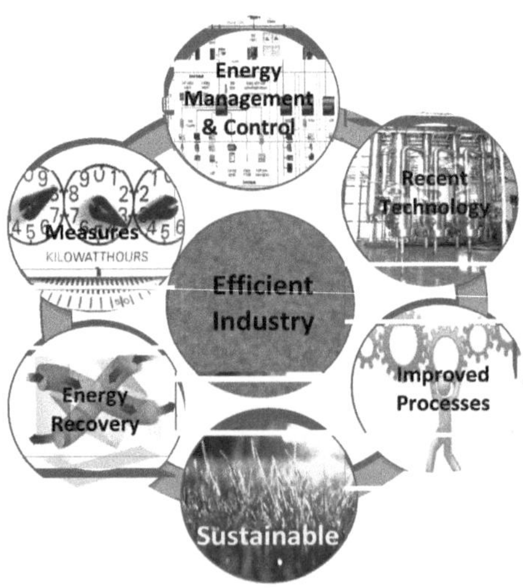

Figure 1.6. *Energy-saving components*

Industrial energy saving is an innovation that consists of modernizing technologies, redesigning and improving the energy use efficiency of processes, constructing energy management and control systems, optimizing power generation, heating and cooling systems, and using the right measures. That is why it should have a value shared with the society.

In several countries, employing an energy manager and starting energy management to control energy use is a legal obligation for manufacturing companies that use more than 1 MWh of energy. Energy Management systems have evolved from pneumatic machines to computer integrated manufacturing. These systems are constructed as an automation system that is composed of a network of microcomputers communicating with a standard protocol. Widely known supervisory control and data acquisition is one of the hardware used on the network. The energy management system can be used to complete the following six steps given by Hatley *et al.* [HAT 05]:

Assessment: this involves the determination of the qualities of energy using factory machines. The analysis includes determination of the existing hardware, communication and standards used among them, mechanical features, theoretical and practical energy use by each of them. A detailed calculation of energy use is compared with the industrial standards.

Plan: integrated with the business plan, the future energy use will be predicted. The tolerances and the standards to be used are decided. There is a need for closing the gaps in the industry. Risk considerations should also be part of the plan.

Design and implementation: this is the area where the manufacturing company needs to take consultation. Experienced engineering companies can differentiate the design of the system network according to the request given by the plans. The basic group of simple scheduling and water sets can only serve a single building. Advanced design differentiates the load schedules of all the machines used in the factory and decides on the level of hot and cold water to be kept according to the environmental humidity. Detailed thermostats and air controls are advanced features that should be considered during the system network's design.

Training and culture: An energy management and control system is not only used for training the operators to use the system. The whole company is aware of the changes caused by energy-efficiency considerations. Awareness training for everybody working in the company is a necessity. This will be the first step in the ecological culture of the company.

Operation: Systems are used in accordance with strategies and plans. Continuous operator awareness about the plan and actual differences will be part of the effective operation. Preventive maintenance of systems and the energy management and control system should not be delayed.

Review: as part of the systems approach, strategies, plans, operational analysis and tolerances should be reviewed at least once a year. This is a system that is in interaction with the weather. That is why in complex systems, at least tolerances and shift plans should be reviewed seasonally.

Assessments for qualified energy management will lead to revisiting the technology for process improvements. It is observed that industrial machines are used longer than their theoretical lifetime by small- and medium-sized enterprises (SMEs) in developed countries and by all industries in developing countries. However, it is well known to manufacturing specialists that less than 70% efficiency of any technology brings more costs than the initial investments. Replacement of technologies and processes should be optimized.

"Energy conservation has often been referred to as the 'fifth fuel', the other four being the so-called primary fuels of coal (solid), oil (liquid), gas and nuclear/hydroelectricity" [BUR 01]. This emphasizes the importance of reducing the amount of energy used and waste heat is an important part of it.

Efficiency concern about the increasing costs has begun in the industry due to leaks in heating processes. The first systematic study of feedback control of steam engines by J.C. Maxwell was performed in 1868. Later, in 1893, A.M. Lyapunov published his first paper on the stability of motion, but his work made an impact on the control theory literature only after 55 years. Thermal recovery is grounded by the improvement of heaters and coolers used in the manufacturing processes. The main purpose is to avoid the energy costs caused by the leak in the valves or transmissions. Heat recovery can be made, or recuperators or waste heat recovery can be used for other purposes.

An example for the benefit of energy recovery is demonstrated by a small exercise in a dye house. If the gases that come out of the dye house are used to heat water, energy use drops by 37%. Further use of heat recovery equipment will reduce the energy consumption further by 50% [HEW 99].

Intensity is seen as a technological performance of the energy consumption efficiency in an industry. It is defined as the ratio of total energy consumption to economic added value of the industry or the total production value.

A detailed study on the evolution of energy intensity performed in Brazil during 1995–2005 by Silva and Guerra shows alternative approaches to intensity measures [AND 09]. In that study, the total energy consumed is given as an average thermal value and production value at constant currency used.

The average total energy is the product of price per energy source at time period t, thermal energy given in Joules by one source in the same time period and the amount of energy used in the same period is the sum of all sources multiplied by the price:

$$E_t^{ther} = \sum_i^N J_i E_i^t \qquad [1.1]$$

where there are N energy types and

E_t^{ther} is the average amount of thermal energy expressed in joules at time t

J_i is the amount of thermal energy in Joules per energy type i

E_{it} is the amount of energy I used at time period t

$$\frac{I_t^{Theil}}{I_{t-1}^{Theil}} = \prod_i^n \left(\frac{E_{it}}{E_{it-1}} \right)^{W_{it}^*}$$

$$W_{it}^* = \frac{[1/2(w_{it} + w_{it-1})w_{it}w_{it-1}]^{1/3}}{\sum_i^N 1/2[1/2(w_{it} + w_{g-1})w_g w_{it-1}]^{1/3}} \qquad [1.2]$$

$$w_{it} = \frac{p_{it}E_{it}}{\sum_i^N p_{it}E_{it}}$$

I_t^{Theil} is the Theil index at time period t (the maximum possible entropy of the data minus the observed entropy)

p_{it} is the price of the source of energy i at period t.

The production value at constant currency is expressed as the vector product of the goods produced and the price vector for the goods at an industrial price index. The researcher claims that the smaller the ratio, the more efficient the energy use in the industry:

$$PV_t = \frac{(Q_t.P_t)}{IP_t} \qquad [1.3]$$

where

Q_t is the vector of goods produced by an industry at time period t

P_t is the vector of price corresponding to the goods products at time period t

IP_t is the price index for a certain industry at time period t.

Hence, the intensity is given as follows:

$$I_t = \frac{E_t^{ther}}{PV_t} \qquad [1.4]$$

A recent study by Proskuryakova and Kovalev differentiates between the intensity and efficiency of energy consumption [PRO 15]. The study shows that the thermodynamic or physical indicator at the lower level is company specific, whereas macroeconomic calculations on the country, industries and sub-industries as the top level give efficiency indications at the country level. Figure 1.7 shows holonomy of energy use that best expresses the concept used by Proskuryokova. The first level is the individual company efficiency that should be considered as part of technological efficiency. The industrial efficiency is the combination of company efficiencies. This leads us to the macroeconomic evaluation of a country that can be expressed as the combination of energy efficiencies in all industries. This should continue to be part of the global governance.

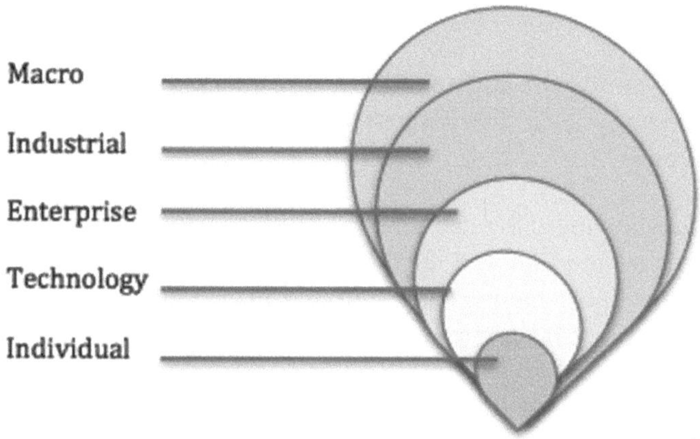

Figure 1.7. *Holonomy of energy efficiency in a country*

Mulder's work on differentiation of intensities provides evidence that energy savings are not only specific to the industrial needs and technologies, but also to the economic state of the country [MUL 15]. Thus, intensity can also be used as a measure of energy-saving culture. A good evidence of this statement was observed in the early 1980s. The European studies during those years showed that with the then current technologies, Danish business was able to save almost three quarters of the energy used, whereas Germans were able to save only half of it [RUM 08]. Rocky Mountain Institute still declare that the majority of the energy use is inefficient, independent of the economy.

Major industrial and technological developments are accomplished by gaining more knowledge about the possibility of improving the efficiency. Malanima pointed out that classical calculations of the production value based on the function of labor, resources and capital are not valid when energy is considered [MAL 10]. Even if energy is taken into the calculations as part of resources, knowledge and technology multipliers are not used. However, energy should be considered as part of industrial technology. He also stated that "The increase in productivity of energy is a consequence of both discoveries of new sources and technologies (macro-inventions) or improvements in the exploitation of those already existing (micro-inventions)".

It can be represented by the intensity ratio with energy expressed in thermal or calorific value.

This is an important statement that considers a wide variety of energy resources currently used. Only a part of energy input is actually transformed into useful energy.

Hidden costs should be considered while calculating the cost of energy saving, but they are ignored when only intensity is taken as a measure. Governments suggest having an integrated energy-saving measure that will consider both industrial improvements and policy effects as well as cultural enhancements.

Energy Saving Company declares that energy savings can be calculated on an annual average basis, which means seasonality or weekend and holiday effects are not required in the calculations due to the randomness nature [DEN 15]. Based on this statement, Deng formulated the energy-saving revenue as a discrete stochastic process for an economic lifetime. Deng's study showed that the larger the investment for the savings, the longer the lifetime expected from all the energy units if both the price and the production value are considered with stochastic multipliers.

A major problem of the process industry is minimizing the cost of intensified energy-saving technologies.

The Canadian Federation of Independent Business made a revelation for sustainable thinking in 1991 [CÔT 95]:

> "*The mistake of the past has been to view the environment and economy as separate cycles connected by a one-way movement of resource from nature to mankind. The ecosystem and the economy are, in fact, very similar in structure and performance. They each require elemental diversity and free-flowing circulation to function properly...The objective is the use of natural resources and develop the return flow of resources from economic activity back to nature.*"

According to Mercier-Laurent, companies and societies should go beyond the reduction of CO_2 emission as decided by COP21. She adds that this is only possible by being vigilant about planet conditions, being curious and aware of what we already have in nature and how to use these resources wisely [MER 15]. “We have to understand things as whole ecosystems before creating improvements or new things dissociated from their natural context”.

“The biosphere consists of living things and the environment – the atmosphere, the land, and the oceans – from which they derive the energy and nutrients needed for life. For instance, plants capture solar energy and use it to convert carbon dioxide, water, and minerals into oxygen and food. Humans and other living things take in oxygen and food and return carbon dioxide and other matter to the system. This cycle repeats itself and everything recycles. Thus, the biosphere can sustain life indefinitely”.

If only we could observe some simple rules of the biosphere, our need for energy will decrease.

Therefore, companies should focus on environmental principles only though eco-designing and recycling. A new creativity culture considering multiple goals should arise, combining economic, environmental and social concerns as well as the balance of related ecosystems.

Among them is switching from relocalization-out to relocalization-in and the use of local talents. It will avoid the transportation of goods that we can produce locally. Another change that may reduce the need for energy is switching from bigger to small and smart, from megacity to well-being, from having more to be happy with little, etc.

The book *Sustainable Industrial Design and Waste Management* published in 2007 indicates that any industrial goal must include preservation and improvement [ELH 07]. Dr El-Haggar mentions the ways to encourage large improvements to the industrial interactions with the environment, giving examples of the sustainability efforts of heavy energy using industries.

Industrial contribution to sustainability can be summarized in different ways, providing economic, environmental and social results.

Clean production: this consists of redesigning the production processes and products, considering the multi-objective optimization to improve production efficiency while reducing emissions and minimizing costs. Decisions made regarding regulations and human interactions with ecosystems come into play. Appropriate modeling involves various time scales and time-dependent uncertainties, where pareto-optimality can be used. This is the easiest step to be taken in manufacturing industries where continuous improvement has become part of daily life. Cleaner production will reduce resource use in addition to improving social health, global heating and economic productivity. This activity is named as remanufacturing and redesigning in manufacturing companies.

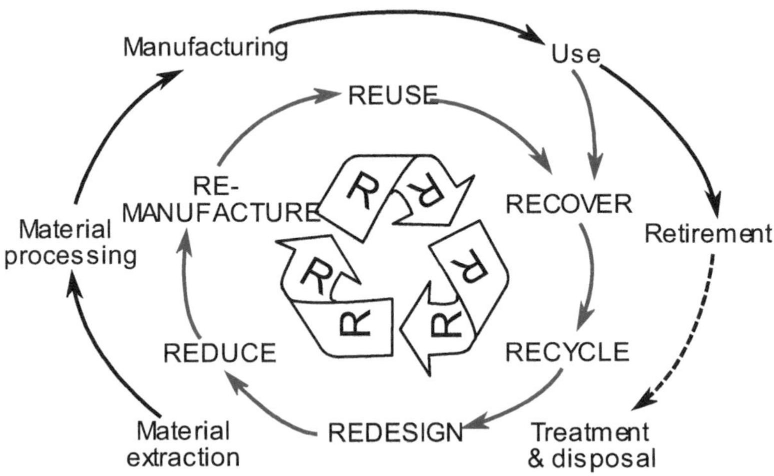

Figure 1.8. *Sustainability in manufacturing expressed as 6R [JAY 10]*

Clean resource utilization: in addition to reducing energy input through efficient processes, the use of renewable energy instead of fossil resources creates a positive environmental effect. With the privatization of power sales and tolerance for micro-generation, there is wide interest in purchasing or self-generation using wind, solar, geothermal or tidal energy. In 2010, 61% of the Forbes Global 2000 companies showed their desire to use renewable energies [COW 10]. Service companies that took part in Cowan's survey stated their contribution to energy saving by using green buildings. Results of this survey will no doubt contribute to resource saving, and the environmental improvement survey no doubt will contribute to resources and environmental improvement.

Waste handling: waste can be reused, recycled or disposed for energy generation. For example, methane gas is a pollutant that escapes through the chimney, but recent technologies allow the recapture and production of steam or even the generation of power. Waste water can be cleaned and reused in the processes. Biological waste such as food and wood can be used to generate energy. Revaluing the waste not only needs investment in technology, but also requires regaining of the costs spent. Manufacturing companies should carefully calculate economic savings when environmental improvements are enhanced [DAI 09].

Recycling: the used product can be recollected if the components can be remanufactured. It is a long flow of the green supply chain that is very much reliant on working with the suppliers and dealers. Supply chains should be designed to integrate with production. Glass, metal products, plastic products and electronic components can be recycled. However, all the recycling processes use more energy. Whereas, recycling is suggested for energy saving. [MER 13].

Ecological innovation: this action involves combining all the above and integrating with users and managers of all sustainable activities. According to the European Innovation act [EUR 06], eco innovation is defined as "*the creation of novel and competitively priced goods, processes, systems, services, and procedures designed to satisfy human needs and provide a better quality of life for all, with a life-cycle minimal use of natural resources*". Corporate Social Responsibility (CSR) was introduced in the 1960s, but brings today recent concerns. CSR is about a company's sense of responsibility towards the community and the environment (both ecological and social) in which it operates. Energy production and use should also consider wise exploitation of locally available resources. The norms (ISO 26000) were elaborated to guide the companies in achieving the objectives of the CSR.

The European Union established the program Energy 2030. "EU countries have agreed on a new 2030 Framework for climate and energy, including EU-wide targets and policy objectives for the period between 2020 and 2030". These targets aim to help the EU achieve a more competitive, secure and sustainable energy system and to meet its long-term 2050 greenhouse gas reduction target. The EU is encouraging private investment in new pipelines, electricity networks, and low-carbon technology. The targets are based on a thorough economic analysis that calculates the costs.

The targets are as follows:

– a 40% cut in greenhouse gas emissions compared with the levels in 1990;

– at least a 27% share of renewable energy consumption;

– at least 27% energy savings compared with the business-as-usual scenario.

Adequate policies are also proposed to help achieve these goals. New attitudes of eco-innovators, open and holistic thinking, and multidisciplinary knowledge are required [MER 11]. This scares companies because of high financial investments. Therefore, small- or medium-sized companies collaborate with others to share the expenses.

Collaboration is mainly seen in industrial parks. Denmark is a very good example of industrial parks [KOR 01]. Unfortunately, this industrial park is not yet evolved enough to use the local renewable resources, but imports coal and oil because of lack of collaboration.

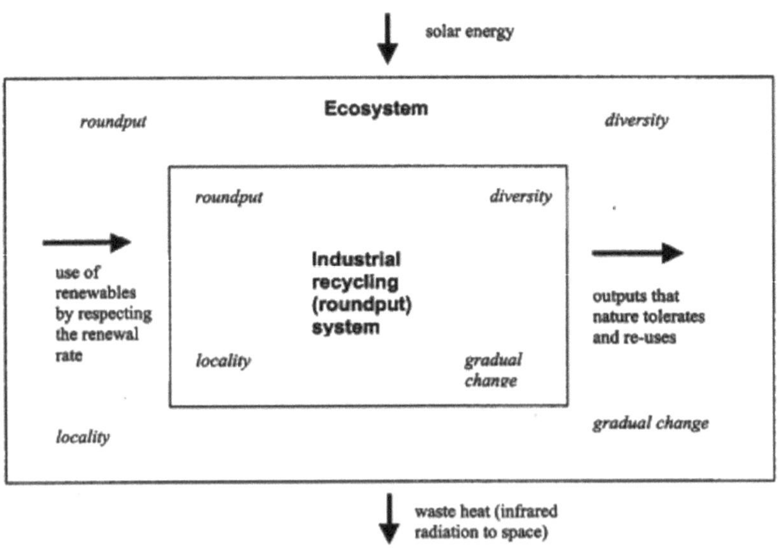

Figure 1.9. *Industrial park with recycling focus [KOR 01]*

A very recent case in Norway shows collaborative carbon capturing between heavy industries located in the same park [NØR 12]. The principle is based on collecting the exhaust output from the steam production and cleaning it before carbon is emitted into the air. This shows the success of collaborations when small-sized companies are not included in the industrial park.

1.4. Impact of technological development

Individual energy use is increased by technological development [MER 15].

Most of the researchers state that the transition to renewable energy use is influenced by climate change. It is certainly true, but the reduction of carbon emissions is not the only reason for the change. We will analyze the impact of technology in parallel with observations on energy innovations. Jacobsen summarizes the following impacts [JAC 01]:

– Technological innovations have opened the room for new energy resources and low-cost supply technologies. Energy generation has become easier, faster and cheaper. Since improvements were made in small baby steps, it was a gradual change that took long time. A good example is solar energy generation widely used in 20–30 years.

– Improvements in technology have caused the innovation of end-user technologies with a focus on energy saving. Energy distribution and measurement has been widely affected as well as the residential consumers.

– Innovation of new production processes has created changes in industrial structure that has indirectly affected the energy demand. More automation, ubiquitous computing, improved accessibility and sharing have increased the demand for energy. However, new energy resources have increased the demand for technology.

– New consumption products have created a paradigm shift in energy demand and consumption. Savings made by more efficient electronic devices might be overcome by future innovations that consume more energy.

The fast technological development enabled the reduction of investments in using renewable resources for energy generation. The liberalization of European energy industries has contributed a lot to the new regimes and policies that encouraged the renewable energy use. Green certificates and carbon trading are good examples of these regimes.

The change in energy use has occurred with respect to the following four results of innovation: transformation, reconfiguration, substitution and alignment:

– *Transformation* is characterized by the gradual adjustment and reorientation of energy use with new objectives like greening. Research investment is modest, and radical approaches hardly exist.

Technology acts as a critical condition for efficient transformation of industrial production. Shao states that the impact of technology on greening can be measured by research intensity and the improvements made in energy use [SHA 11]. Industry has been focusing on the use of fast developing technologies mainly to minimize the amount of resources used. From heating, ventilation and air conditioning applications to the most recent zero energy concepts, energy generation technologies have been the center of interest. Building refurbishments, steam and waste recoveries, and machine replacements are all based on efficiency performances. The use of batteries for power storage has made it possible to use hybrid resources of renewable energies and grids. It is possible not only due to the materials used (e.g. lithium), but also because of the technology of charging and recharging that gives longer life to batteries, making the transformation of resources possible.

The Internet of things and cyber systems have led to Industry 4.0 where the real transformation of industrial energy use can be observed [LEI 16], and the multi-agent applications in energy use will certainly integrate the resource use.

– *Reconfiguration* is defined by niche innovations that cause radical changes using state-of-the-art technology. Hence, new components or technologies are presented for energy use. Cogeneration is certainly one of them since the technology developed to create heat using different resources only started after the economic crisis that occurred in the late 1970s.

Self-sufficient energy generation for a single home is a big change for residential energy use. Regional grids are a reconfiguration, which is possible only due to new operational control technologies including knowledge systems.

Reconfigurations are also triggered by new algorithms and methods that are developed. Performance measures and indices are redefined by using state-of-the-art methods as in rearranging the power distribution system to optimize the energy delivered [IMR 14].

– *Substitution* provides opportunities for niche innovations when problems exist in the current system. Radical innovations can use these opportunities to enter increasingly big markets [MER 15].

A continuous improvement in line with technological development needs investment in R&D. It is observed that investments are dependent on changes in price. International environmental regulations further affect the market models with multi-country, multi-company, multi-technology energy generations.

R&D operations result in industrial ecosystems for better greening where resources to maintain technological innovation and resources to maintain innovations in energy generation and consumption are shared.

– *Alignment* is required among the policies changed by new technologies. The legislation is modified due to global interactions and easy accessibility proposed by state-of-the-art technologies. Uncertainty of new technologies is tested and, finally, one option is chosen for radical change.

Waste collection and reuse for energy generation is an alignment. Methane or steam collection used in cogeneration /tri-generation facilities depends largely on technological developments. Industrial buildings and logistics are also improved by an alignment of energy technologies.

Manufacturing use of energy accounts for 35% of global energy use. Industrial energy use involves running the production processes continuously. Any break is a cost issue. The second field is a smaller amount of use for the heating and cooling of industrial buildings. However, both inbound logistics (transportation and storage of the materials and products in the plant) or outbound logistics (to deliver the products to the customers and use warehouses for distribution) are heavy energy users. A recently growing field of industrial energy use is the operations of waste collection and reuse as in the case of steam and methane gas.

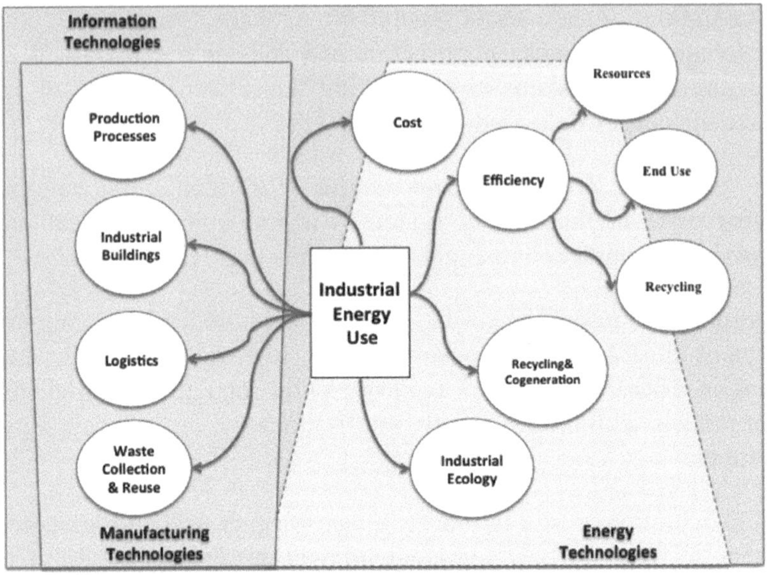

Figure 1.10. *Technology interaction in industrial energy use*

As shown in Figure 1.10, during the use of energy in the industry, there are always cost and efficiency concerns. Efficiency capitulates the energy resources used and the recycling of the waste, in addition to the savings by end users. Self energy generation is growing through different industries using co-generation/ tri-generation techniques.

An excess heating system in a tri-generation system can be predicted by using equations [1.5] and [1.6], with c being the specific heat of water kJ/kg °C and m the flow rate [CHE 16]:

$$Q_t = c.m_j.\Delta T_t \qquad [1.5]$$

where m_j is the water flow rate in the jacket and ΔT_t is the temperature change.

$$Q_t = c.m_j.\Delta T \qquad [1.6]$$

where m_{ch} is the water flow rate in the chiller and ΔT_{ch} is the temperature change in the chiller.

Information systems have always been essential for industry since the very first approach to inventory management. Now, it is a basic necessity as energy use should be observed and controlled dynamically and in real time. Most recent studies have shown that information technologies strongly affect the optimum lifecycle of industrial energy use [CER 16].

Energy efficiency has always been the golden keyword in the renewal of technologies, whether it is energy-efficient manufacturing technologies or energy technologies. A recent article has analyzed energy-efficient technologies to determine whether they are saving energy. It is observed that energy-efficient manufacturing technologies use less energy themselves, but processes with more computer power require more electricity [PEL 12].

Inbound and outbound logistics should be considered as energy users and more focus is given to the energy use of supply chains, as suggested by Wang [WAN 16].

Most recently, ecological apprehension has become a part of energy use objectives. As the energy culture in the industry develops, dependency on energy technologies multiplies. Moreover, state-of-the-art manufacturing technologies are integrated with energy technologies.

When the input and output of all the industries are considered, the total output of each sector can be defined as the output from that sector (i) to the others (j) and the ultimate demand of the sector i is given by:

$$O_i = \sum_{j=1}^{n} X_{ij} + D_i \qquad [1.7]$$

Then, the energy required is the unit energy intensity E and the output together:

$$e = E.(I - \frac{X_{ij}}{X_j})^{-1}.D \qquad [1.8]$$

where *D* is the demand matrix of all industries and *I* is the identity matrix and intensity is the ratio given

1.5. Industrial leaning

Energy analysis is studied in energy supplies and utilities for use in transport, industry and buildings. In 2005, more than 50% of energy was used by industry; with all the efficiency efforts, it drops to 38%. The most energy-intensive industries are iron/steel/ aluminum production, petroleum refining, chemical products, glass, pulp and paper, and non-metallic mineral products such as cement, lime, limestone and soda. Less energy-intensive industries include the manufacture or assembly of automobiles, appliances, electronics, textiles, food and beverages, and other products. In the USA, only petroleum use in the industrial sector increased by 24% from 1985 to 2002. Energy use in manufacturing industries continues to be significantly higher than that in industries such as agriculture, forestry, fisheries, mining and construction.

With the increasing price of prime sources, the progress in energy efficiency is triggered. The global competitiveness of heavy energy using manufacturing companies is based on price-induced efficiency in addition to governmental incentives. Hence, the major target of efficiency studies in the 20th Century was to reduce the total costs of energy investment and operations. Productivity benefits were calculated based on process assumptions incorporated with the active technologies in the industry. Ernst *et al.* [ERN 78] classify energy-efficient technologies into three main groups, namely process improvements, efficiency gains and conversion (energy generation and efficient material use). Since then, this classification has been used. Worrel did not consider the technologies used for sustainability and clean environment, which became an important subject of interest later [WOR 03]. Even in 2009, the International Energy Agency studied the industrial focus for clean air as a future transition. Worrell calculated the value of productivity benefits based on bottom-up energy conservation scenarios. Conservation supply curves were generated to evaluate the energy investments with costs on an energy unit basis. These curves also indicated the amount of energy to be supplied under certain prices, considering the costs of energy conservation.

Although industries still use 80% of coal, 43% of natural gas and 8.5% of oil, industrial share in electricity use was reduced from 56% in 1979 to 42.3% in 2012 [INT 14b]. A 75–122 billion kWh saving is a strong possibility by using new manufacturing techniques as indicated in America's Energy Future [NAT 09]:

– separation processes with improvements in thermal efficiencies;

– materials for resistance to corrosion, degradation and deformation;

– nano-materials for structural applications;

– technologies reducing natural gas;

– state-of-the-art boilers;

– electricity-saving electric motors;

– cost-effective energy control systems;

– new generation of sensors with inferential controls and distributed intelligence.

However, a thorough assessment would require the analysis of the options and qualitative management of investments.

Industries have applied learning-by-doing, learning-by-using and learning-by-interaction to apply and improve the energy technology use. They have learnt to follow industry-specific applicability in parallel with the possibilities to increase economic advantages. This learning brought the energy management applications.

Energy management can be defined as doing the best to reduce energy consumption in addition to energy conversion without any leak. For these two purposes, all energy consumption levels and variations must be known for each product line or item of equipment. From these data, we can determine how much energy is consumed uselessly and how it can be reduced. Lean manufacturing techniques can be used to collect these data. Future value maps showing improvements can be motivating for employees [KES 12]. Improvements could target optimum fuel/air ratio combustion (e.g. furnace or boiler operation adjustment), prevent the leakage of steam, oil, water, compressed air, etc., strengthen heat insulation (e.g. furnace, heater, dryer, piping, tank, etc.) and install simple measuring devices. Introduction of new processes, changes in automation systems, or installation of heat recovery systems or the inverter can also be planned for the future.

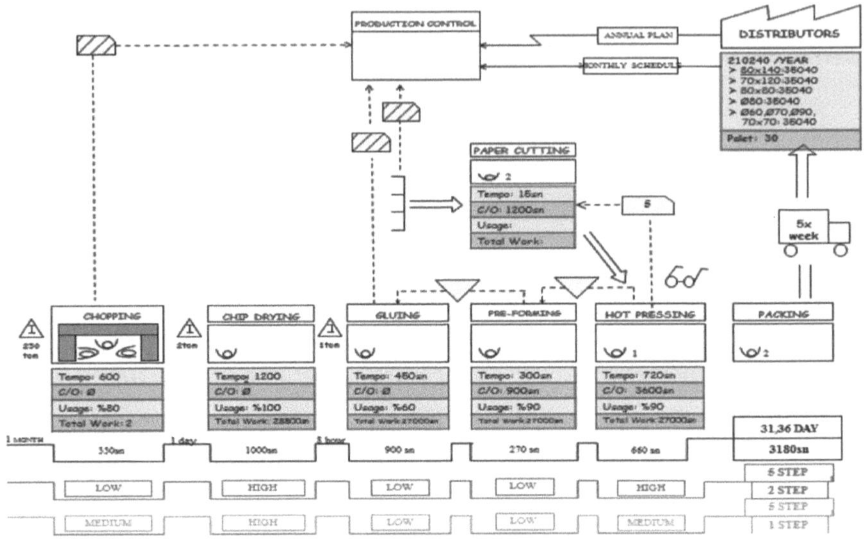

Figure 1.11. *Sample of a value stream map for energy use [KES 12]*

In some cases, new processes may bring much benefit or even better productivity or quality of products. Examples of these cases are as follows: cogeneration or tri-generation systems; heat recovery system; suspension pre-heater; counter-flow type continuous digesting system for pulp industry; jet-type dyeing machine in the textile industry; combined cycle power generation system; regenerative burner system for industrial furnace. The important decision for improvement is taken by calculating the return on investment (ROI). Some heat exchangers saving 3 kWh per day can pay the initial investment back in 2.4 years; moreover, an inverter that saves almost 30% of the power used for the cooling pump can have a ROI in almost 3 years.

Studies on industrial energy-efficiency improvement in the USA considered 14–17% saving potential for 2020 [NRC 09]. Scenarios for clean energy were suggested by Letcher with an expectation of 16.6% reduction in industrial consumption [LET 08]. In contrast, the McKinsey Global report published in 2009 leaves the estimates of industrial energy consumption reduction at 10% [GRA 09]. According to the UK government, 196 TWh of energy (equal to 22 power stations) can be saved in the UK until 2020, of which almost 50% can be achieved by industry [CAH 12]. The majority of the EU members agreed to a policy of 20% reduction of energy consumption in 2020 [YUN 11].

Industrial energy consumption has seen a significant reduction in Japan, from 12% in 1975 to 6% in 2010. Similarly, Chinese manufacturing companies have reduced their consumption by almost 60% since 1980 [ZHA 14]. The estimated targets could not be achieved either in Europe or in the USA or Japan. A detailed study of efficiency drivers shows that audit-based energy-efficiency policies are less effective than the company-based policies [THO 09]. Obviously, individual policies vary according to the size of the company as well as the business culture.

Self-energy production using cogeneration or tri-generation is a good example of company-based policies, besides steam or gas waste recovery. Energy-efficiency policies may also be evaluated by the sustainability of products and services, decision-making processes, management styles and innovation speed. Lack of skilled personnel is an especially important challenge for SMEs. The invisibility of energy-efficiency measures and the difficulty in quantifying the impacts, lack of inclusion of external costs of energy production and use in the price of energy, and slow diffusion of innovative technology into markets are considered as barriers. Voluntary industrial agreements in Japan and Germany lead to company-based policies.

To define the most effective company-based policies, the company should not have an organization that works top-down, and instead define the standards and values for energy use; the role of energy for the business should be known and communicated; an effective management should influence the mindset of all the workers; the company as a whole should understand and anticipate that efficient energy use is a successful business for tomorrow.

Recovering energy waste has become an important focus point in the sustainability of manufacturing companies in this century. For example, in a heavy energy-consuming industry, cement production, residual energy in the fumes of the furnaces is recovered by using new technologies in order to reduce the cost of products. However, in the paper and pulp industry the focus has been on recovering steam and avoiding the use of large volumes of cold water. In chemical industries, the recovery of heat energy from the combusted waste gas has reduced the production of toxic gases.

Manufacturing sites have tried to become autonomous in energy production, maximizing the use of available energy. Harvesting both thermal and kinetic energy for electrical energy to power the manufacturing devices has been encouraged by the improvements in storage technologies [GAR 12]. Prabhu gives a new definition of energy-aware manufacturing as follows [PRA 15b]:

- real-time or forecasted manufacturing supply or needs;
- green energy or energy cogeneration in the grid;
- real-time measurement or forecasted energy consumption;
- energy-efficiency measures;
- products, resources or process redesign as energy-aware;
- recycle energy wastes;
- energy savings.

1.6. Environmental constraints

The renewed awakening about climate changes and planet conditions pushed politics to action. The Rio summit in 1992 was the first action, but the decisions taken were not easy to respect because of the global focus on business only. Following the Kofi Annan proposal made in Johannesburg in 2013, participating governments agreed on the targets. During COP21 in 2015 they signed an agreement on reducing CO_2 emissions. We will provide the details in the following chapters.

As with other companies in all fields, the energy producers have to respect the norms of CSR, such as ISO 26000. CSR focuses on the balance between three components, namely economic, environmental and social, as shown in Figure 1.12.

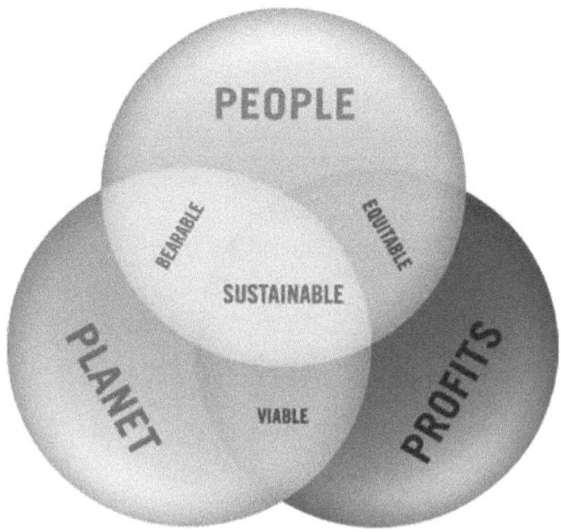

Figure 1.12. *Basic principle of CSR [PAS 13]*

Each company adapts its CSR program as a function of its activity. Many focus on environmental actions only, for example ENGIE, a French national energy company (see Figure 1.13).

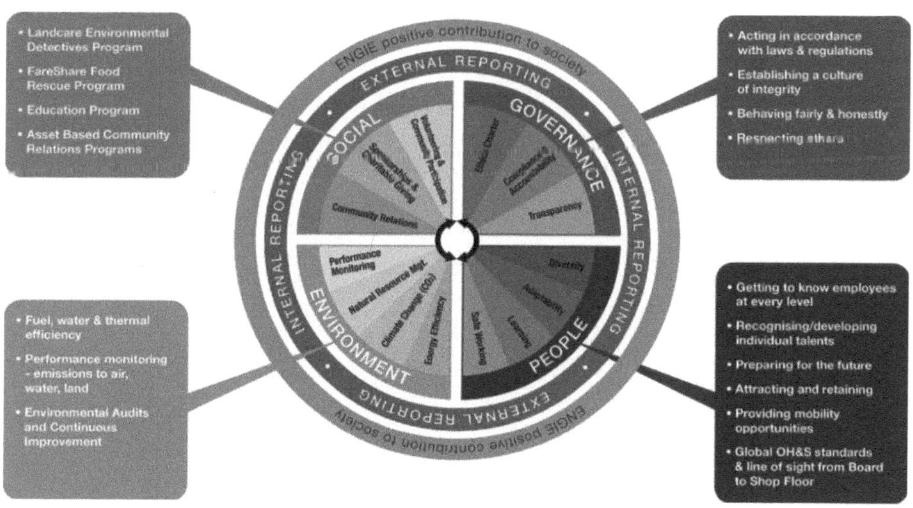

Figure 1.13. *CSR program of ENGIE [GDF 16]*

SMEs need to elaborate on adequate strategy and programs considering the balance of the three components.

Involvement of any company depends on the allocation of resources for these activities.

It is mainly about awakening individuals and companies and undertaking actions to stop wasting energy. This is not easy to realize in a world where quick business is preferred to long-term well-being.

2

Technology for Energy

Energy cannot be generated without using natural resources. The method used, whether we harness energy from the Sun, dig up the fossilized form, or use a new material like fuel cells, is just a detail. A Möbius strip is a good representation for the intricate relationships between humans and energy. Energy generation should not harm nature. It can only be possible if humans are inspired by nature.

After years of energy-saving experiences, we now use renewable energy and fossil energy in a hybridized manner. Prime energies of coal, fuel and gas were discovered by mining, but hydro, wind and solar energy by conversion. That is why we start the lifecycle of energy from energy generation. Resources can be found in depths of the Earth, in the atmosphere and even in waste; there is a transformation so that they can be used as electrical energy, for heating or cooling. Energy generation is liberalized over the globe, but nuclear energy plants are run or closely controlled by the state to guarantee the safety.

Energy transition is still run by the state in the majority of the developed countries due to security. It is very important that the infrastructure of national grids is robust and secure. Transformers are installed in grids all over the world, with China recently becoming the best market for transformers. The improvements in high-power density technologies aim to increase capacity with intelligent approaches while improving environmental friendliness. Energy storage is enhanced in parallel with the hybrid use of renewables and has been the reason to design transmission systems.

Energy distribution has become even more important after the development of smart grids. Sales level is an important stage before the use of energy, because the majority of the markets are liberalized.

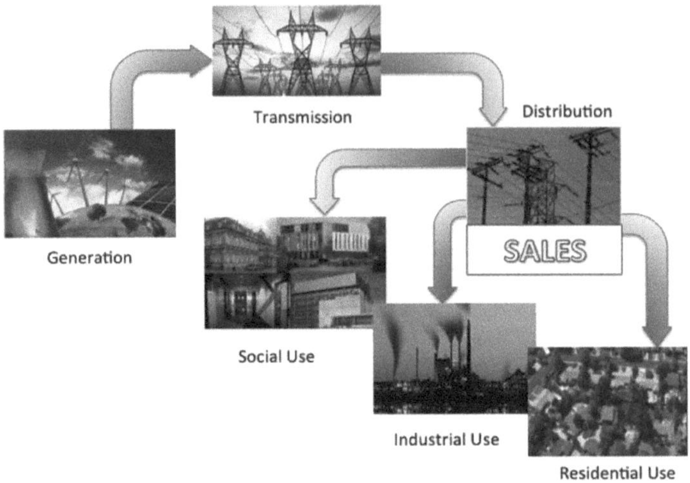

Figure 2.1. *Energy lifecycle*

Energy users can be social, industrial and commercial, or individual as shown in Figure 2.1. Individual users are connected to the national grid in most of the countries. Residents wishing to become autonomous use locally available alternative energies. Hitachi and some other companies provide small (maximum 1 kWh) hybrid generators which can exploit water, sun and wind. Social use represents schools, hospitals and governmental buildings, which are considered as social costs in some countries and taken into account in commercial use for the others.

This book focuses on industrial and business use of energy; however, most of the new technologies support the whole energy lifecycle. Because information and communication technologies (ICT) are everywhere, we call the current period a Digital Era.

Technologies, applications and some perspectives of the Digital Era are shown in Figure 2.2. Big data is becoming very popular, because we accumulated a huge amount of data which are the raw materials of ICT. The current trend is to try various knowledge discovery techniques, such as analytics, induction, genetic algorithms, neural networks and others, to find relevant information in this hay stack. Some companies, cities and organizations are dreaming about making business from accumulated data.

Mobile applications are also quickly growing to offer new solutions such as remote monitoring of energy at home and many other useful applications.

Internet services related to Web 2.0 include social networks that are a part of knowledge flow in some companies if used wisely. Collaborative spaces, e-learning, e-business and e-health applications, and search engines are becoming semantic, with a growing field of connected objects and on-cloud computing facilitating the use of distant server power.

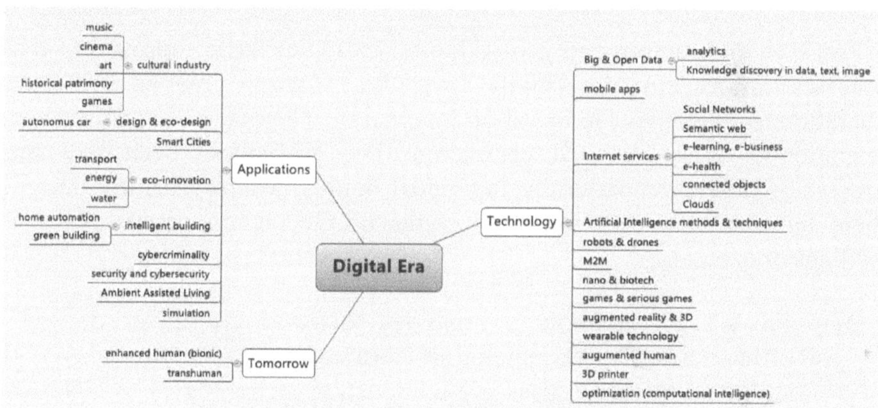

Figure 2.2. *Overview of technologies that can be used in the energy lifecycle [MER 15]*

Artificial intelligence techniques are presented in detail in Chapter 3; their challenge is to help the user in decision-making, knowledge discovery, managing complexity, problem solving, etc.

Machine-to-machine communication technology enters the energy field for automated processing of data related to various processes.

Nanotechnologies and biotechnologies are also growing quickly in terms of applications in many fields, such as food, cosmetics, textile, construction materials, agriculture and in the medical domain.

Augmented reality and 3D have been used for years, for simulation, design diagnosis, training and games. The augmented human was initially developed to help people with disabilities recover some life functions.

Wearable technology is all around us: bracelets, glasses, but also sensors embedded in textiles have become more and more popular.

These technologies open an infinite spectrum of applications in all fields, including energy. It is up to us to be selective and focus on those that really help us to perform our professional or extra-professional activities.

2.1. Technology for the energy generation

Prime energy mining and generation technologies have improved to make processes more environmentally friendly, whereas renewable energy generation technologies are based on innovation. The major change in energy generation is the distributed generation. The grid might even buy small amounts of power generated by individual houses. Micro-grids are designed where the electrical cars store the extras of the energy generated by the buildings or tribunes.

In prime energy, technology improves to avoid the emissions and to increase efficiencies. Power generation from coal, which even today is the most widely used resource, uses pre-combustion technologies to clean coal and post-combustion technologies to clean carbon emissions. Burners, furnaces and fuel staging are used by these technologies to create cleaner energy [SEN 10].

A major improvement in natural gas is combined-cycle gas turbines, allowing for self-energy production in industry and the generation of power and steam. Gas turbines have been improved since 1920 to be more efficient and less disruptive.

In recent years, there has been significant growth in the installed base of fast-start and rapid response gas-fired technology, which minimizes emissions and fuel consumption during start-up and load-following operations. Gas turbine manufacturers have invested heavily in fast-start combined cycle plant designs, which have the potential to integrate well with renewable resources. As a result, combined cycle gas turbine (CCGT) plants now offer significant advantages compared with nuclear and coal. Shorter start-up times, high start reliability, lower start-up costs, daily multi-start up and shutdown, high ramping rates and extended emission compliance at partial loads are included in its benefits, as well as comparatively lower installation cost. CCGT improvements enable regulated power generation that compensates for the fluctuations in renewable resources [NAT 09].

Energy-saving equipment such as packed towers and tray towers are used. In general, the amount of heat absorbed is controlled by the mass of liquid that passes through the device [SCH 13]. Heat recovery boilers, steam regenerators, are part of any chemical plant, refinery system or chemical processing industries [GAN 02].

The equipment is summarized in the following list provided by Mozzo in the Encyclopaedia of Energy Engineering and Technology [MOZ 07]:

– *Waste heat steam recovery*: it is either a water-tube or fire-tube boiler that uses the hot waste heat gas stream to heat boiler feed water to generate either low-pressure or high-pressure steam. Some boilers will be sold without a burner package.

– *Recuperator*: it is an air-to-air heat exchanger. The transferred heat from the waste heat output can be used anywhere as the pre-heated air. One of its uses may be to recover heat from a boiler condensate blown down on the shell side to heat up boiler feed water on the tube side.

– *Fine-tube heat exchanger*: this type of heat exchanger uses air (usually the waste heat stream) blowing across finned coils that contain water. The exhaust gases are dirty.

– *Plate and frame heat exchanger*: this piece of equipment consists of two frames sandwiching thin plates. It is usually used with fluids of relatively low temperature, which flow through alternate plates. Since the plates are thin, heat transfer is usually good; however, both fluid streams need to be relatively clean.

– *Heat wheels*: they are useful for low temperature applications. The two ducts, outside and exhaust air, must be adjacent to one another so that the wheel turns through both ducts. The wheel will turn at low revolutions per minute collecting the waste heat or cooling, depending on the season.

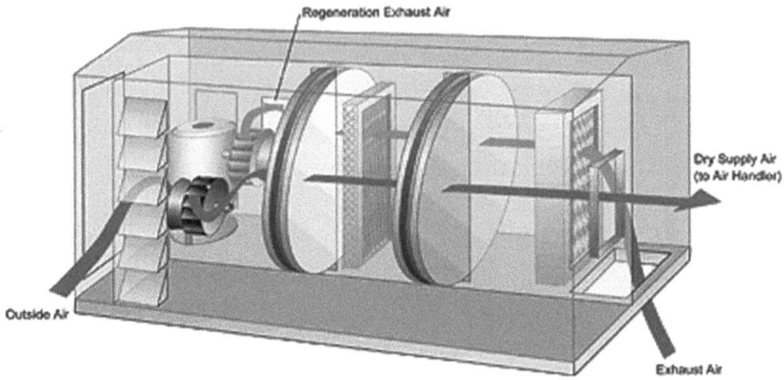

Figure 2.3. *A typical heat wheel [MOZ 07]*

– *Heat pipes*: this equipment consists of a pipe heat exchanger with the interior containing a coolant. The coolant is alternately vaporizing and condensing between an exhaust and outside air stream, exchanging cooling and heating between the two air streams.

– *Run-around coils*: they are similar to heat pipes in operation, cool or heat exhaust and outside air streams. Their typical construction is a finned water coil with air blown across the coil. The advantage of the run-around coil is that the exhaust and outside air ducts can be physically separated. It needs a pump and a water/glycol solution that must be subject to freezing conditions.

The demand for heating grows with the population. But the demand for power is growing as the technology improves and the need for knowledge increases. Because of the environmental dimension of energy production and use, the importance of conducting an energy reappraisal grows in proportion to the world population. However, uncertainties of environmental interactions and the cost of modifications in energy systems are pending the reappraisals.

One opportunity for technological developments was raised with the energy price shocks of the 1970s. The way in which severe shocks to the system can create technological opportunities is illustrated in the history of energy systems. Even energy experts invoked a variety of conservation measures. Based on information collected in the late 1970s, the possible steps could be as follows [HEL 90]:

– optimal localities for heat generation should be found to avoid the heat loss in transmission;

– climate of the agriculture should be created with the effect of heated sera;

– local climate control should be made with long-term weather;

– energy production and consumption should be analyzed with consequences.

After having applied all the suggested improvements, the general heating system, which is still used today, can be achieved, as defined in Figure 2.4. It consists of six subsystems including fuel energy delivery, power generation, power distribution, heat generation, heat distribution and the end-users' subsystem. The function of each subsystem is quite different in nature. The heat supply is indispensable for the system and therefore called the first-class subsystem. Second-class subsystems are those that function to reduce the fuel and/or monetary consumption for the unity end product, although they themselves expend a certain amount of fuel and monetary costs. All types of heating system can be represented by the generalized heating system, with the probable exception of some second-class subsystems. The "hot end" consists of the fuel energy delivery, power generation and power distribution subsystems, while the heat distribution subsystem and the end-users' subsystem are components of the "cold end" of the generalized heating system.

There are a considerable number of empirical analyses of resources, costs and technological options of energy. Paul L. Joskow from MIT gave a thorough analysis of 25 years after the Ford study in 1979. He interpreted nine policies suggested by Ford as follows [JOS 03]:

– deregulate oil and gas producers;

– make consumer electricity prices reflect real costs;

– use science and technology to generate basic options;

– adopt new cost-efficient ways to clean the air;

– prepare a disruption in global oil markets by developing stockpiles;

– continue efforts to reduce problems with nuclear energy;

– make industry use coal with the knowledge of carbon dioxide problem;

– pursue conservation as a new energy source;

– remove impediments to the use of solar energy.

Type of Heating System	System Arrangement	Legend	
boiler	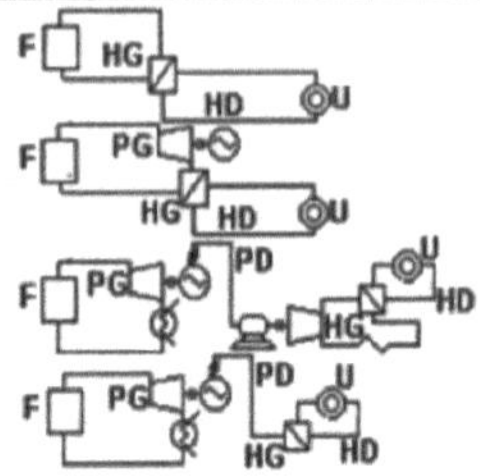	F	Fuel Energy Delivery Subsystem
CHP		PG	Power Generation Subsystem
Heat Pump		PD	Power Distribution Subsystem
Electric Heating		HG	Heat Generation Subsystem
		HD	Heat Distribution Subsystem
		U	Users' Subsystem

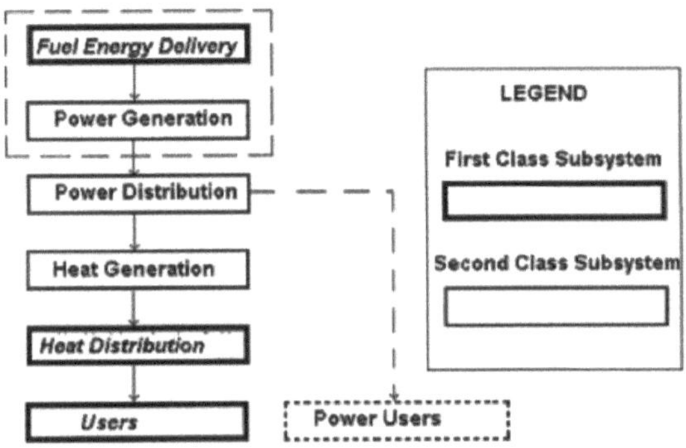

Figure 2.4. *Current heating systems [TOW 14]*

Technologies for energy generation can be classified as fossil fuel conversion technologies, renewable electricity generation, renewable heat generation, biofuel generation technologies and hybrid resource utilization techniques. Since industries are inclined to generate energy both to recycle the steam and gases and to respond to self-energy demand within control, the cogeneration and tri-generation technologies are of high interest today.

2.1.1. *Cogeneration and tri-generation systems*

Conventional power generation, especially electricity generation, using fossil fuels is based on a burning process. The heat energy is sequentially transformed into mechanical and then into electrical energy. Through this transformation process, there is room for waste heat. The simplest way of regaining the wasted energy is to utilize it for heating. Therein comes "combined heat and power" as an effective power supply method. In addition to energy savings and emission reductions, there will be a cost reduction. The UK claims £176 million annual cost saving on the energy production and the reduction of 2.1 million tons of CO_2 emission following the installation of 1 million micro-cogeneration units in the UK [UNA 15]. Currently, combined heat and power generation systems (CHP) are not only used by the industry, but also for responding to the residential sites, especially in colder climates.

Tri-generation systems are considered as an alternative for warmer climates. A typical tri-generation system involves the extension of a CHP plant by coupling a prime mover to thermally driven equipment to produce cooling. By this extension, the waste energy from the prime mover can be used to drive both heating and cooling systems. A combined cooling, heating and power (CCHP) system can achieve up to 50% greater system efficiency compared with a CHP plant of the same size. A simple tri-generation system consists of four main units, as shown in Figure 2.5. The main part of the system is the power generation unit, also known as the prime mover, which produces mechanical power. The second part is the electrical generator that uses the produced mechanical energy. The remaining two parts are heating and cooling units, which can be fed by waste heat totally or partially. Waste heat stems from exhaust gases, lubrication oil and jacket water.

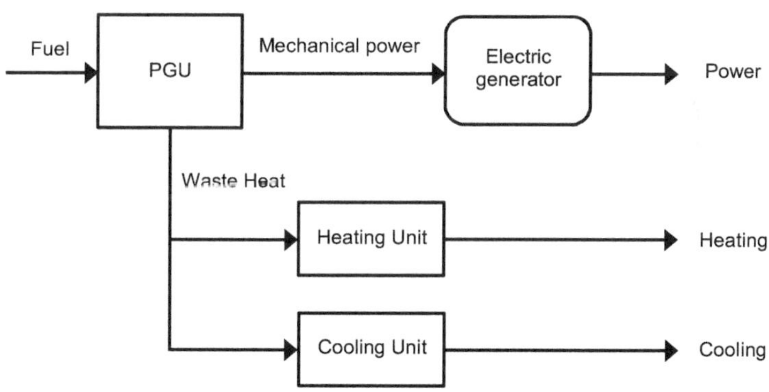

Figure 2.5. *Typical tri-generation system [UNA 15]*

Majority of the tri-generation systems are grid connected, which means that electrical energy exchange is possible between the CCHP system and the electrical grid. As a main part of a CCHP system, a prime mover normally produces shaft work, so it is always being used to couple with an electricity generator. As prime movers, alternatives are internal combustion engines, gas turbines, steam turbines, micro-turbines, Stirling engines, Rankine engines and fuel cells. Of these alternatives, Stirling-based units, organic Rankine cycle-based systems and fuel-cell-driven technologies are still in the research and development stage with limited commercial applications [JRA 14]. The most critical step for designing a CCHP system is selecting the most appropriate prime mover. Recently, wind power and solar panels are also used as prime movers.

Another important part of a tri-generation plant is the thermally driven refrigeration system, in which the rejected heat from the prime mover is used to fulfill the cooling demand totally or partially. The main alternative cooling technologies used in tri-generation systems are absorption cooling, adsorption cooling and desiccant humidifier. Absorption cooling is preferred in industrial cooling or huge spaces, because of confidence to several researchers. Some specifications such as low noise, solution pump free, corrosion and crystallization trouble free, and adsorption chillers for small volumes, might be suitable for small-scale systems.

An extension to tri-generation systems is possible by producing more outputs such as hot water, potable water, hydrogen or some other chemical products. The extended system is called poly-generation or multi-generation.

2.1.2. *Renewable energy technologies*

Growing concerns for our planet have influenced political decisions to support the renewable energy development. In developed countries, efforts have been made to make renewable investment cheaper. The renewable policies have to be consistent with national objectives as well as global objectives [CHO 11]. Renewable or hybrid energy use is no longer implemented as a technological project, but as a strategic obligation. Exemplary countries such as Denmark and Germany decided to cover 100% of power demand using renewable energies.

Among the renewable sources of energy, some are well known, but not sufficiently explored:

– *Hydroelectric power* has been one of the key resources of industrialization. Dams constructed at the side of rivers collect water and turn turbines to generate power without carbon emission. The International Energy Agency (IEA) reports that, currently, 14% of all the energy is produced by hydropower and will be doubled in 2050 [INT 10]. But this renewable resource is damaging to the environment and therefore there are endless oppositions. Water wheels allow the self-energy generation for residences at the water site. The balance between energy generation and local natural ecosystem should still be considered.

– *Wind power* dates as far back as 500 AD in Persia [WIN 16]. However, tower mills were first constructed by the Dutch people in 1390 [CLA 14]. Today, 3% of the global energy demand is responded by wind sources. It is a growing industry and thus wind energy is still creating new job opportunities. Wind turbines are based on the principle that the wind turns two or three propeller-like blades around a rotor. The rotor is connected to the main shaft that spins to create electrical energy. Currently, there are two main kinds of turbine: horizontal axis and vertical axis turbines [DOE 16]. Offshore wind turbines and small- or medium-sized turbines are produced. A wind turbine has 20–25 years of lifetime. All the turbines can run either connected to the grid or stand-alone. At the end of 2015, 44.2% of the European energy investment was in wind power. Currently, 11.4% of European electricity consumption is provided by wind energy. China is just starting the greening process by investing in wind power.

– *Solar energy* exists in nature. Half an hour of solar radiation on the Earth's surface can support a year of global power consumption. There are two technologies that directly convert electromagnetic energy from the Sun into useful energy: solar photovoltaic (PV) and solar thermal-concentrated solar power (CSP). Photovoltaic systems capture sunlight and radiation with silicon cells (discovered by Sharp Corporation from Japan, in 1963) and convert light into electricity. Photovoltaic reaction, in which radiated light excites the electrons of a semiconductor, was first observed by the French physicist Alexandre Edmond Becquerel in 1831. PV technologies use either silicon crystals or thin films. Silicon crystal is commonly used because of its higher efficiency. However, the expenses of refining the silicon are high, which is, why more research is carried out on thin films. Germany has invested a lot in solar energy, and they continue to create more new efficient PV techniques at lower costs. The biggest PV park in the world is Longyangsya Dam in China, generating 850 MW of energy, which was unveiled in 2015. Previously, two solar parks in California, USA were leading in size: Solar Star (579 MW) and Topaz Solar Farm (550 MW). The biggest solar farm in Europe is in France, Cestas Solar Farm, which went live on October 2015, generating 300 MW power. Solar energy will be the key element of flexible energy generation [DOE 16].

– *Wave energy* is the movement of water in oceans that allows the use of water currents as a renewable resource of energy. The difference between the high and low tides have to be 16 m for efficient energy generation. Tidal turbines and fences are used, but dams are also part of tidal energy. Fences are used in currents between the islands. Tidal turbines are smaller than wind turbines and produce more electricity. Waves are also used with the wind blowing on the surface of oceans. An increase in using the ocean saves almost 20–30% cost [DOE 16].

– *Geothermal energy* uses the heat stored in the Earth. It has been used for heating since the Roman Empire. Hot water near the surface can still be used for greenhouses, fisheries, industries and district heating. Heat pumps are used to dig the hot water deep in the Earth. Finding 40°C–150°C might be deeper than expected [DOE 16]. Magmatic arcs formed by several volcanoes exist mainly in the west coasts of Latin America. Recently, devices such as heat pumps and geo exchangers have developed technologically to provide residential heating in cold climates. Denmark is leading in that research. International Geothermal Associates (IGA) reports 27 countries that have electricity-generating installation using geothermal energy. The majority of geothermal energy use is less than 120°C. However, enhanced technologies (engineered geothermal systems) are used for deep systems. Deep geothermal systems can be utilized for electricity, heating, or combined heat and power applications [LUN 16].

– *Biomass* is the oldest renewable energy source used since it includes wood burning, or, in poor countrysides, the burning of animal waste. It is a chemical process that converts plant sugars into gases, which is then burned as ethanol or used to generate electricity [CLA 14]. Wood chips, bamboo, sweet sorghum, tall fescue, kochia, wheatgrass, rye grass, weeds, grape sludge, almond hulls are some of the biomass sources. Wood processing residue, municipal waste and animal waste are also biomass resources. Biodiesel and ethanol are bio fuels, that is, the oil produced using biomass that is not renewable. The power is generated by either direct combustion (steam turning a turbine), co-firing (use of some fuel), gasification (chemical process to produce gas burned in boilers or turbines) or anaerobic digestion (bacterial decomposition) techniques.

All these resources are usually combined to provide continuous production of energy in time [WOZ 06]. The biggest problem with renewable energies remains their storage.

2.1.3. *Energy storage technologies*

Energy storage technologies have been improved with the wider acceptance and use of renewable technologies. These technologies can be summarized into three classes, namely batteries, magnetic storage and flywheels, as shown in Figure 2.6.

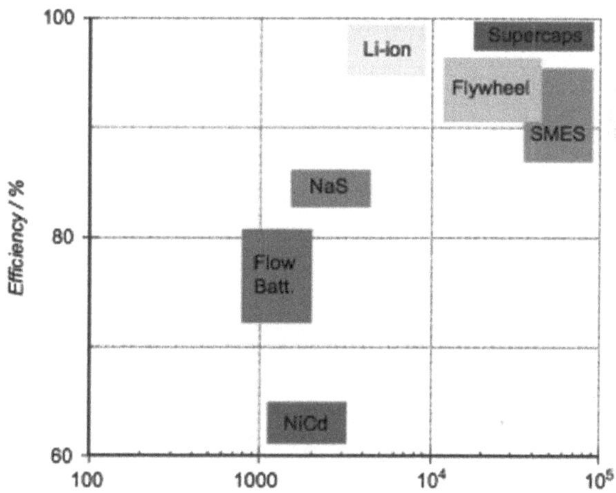

Figure 2.6. *Lifecycle of energy storage technologies [HAL 08]*

Batteries are analyzed together with supercapacitors. They have high-energy density but low-power density storage media, whereas supercapacitors are just the opposite. That is why supercapacitors are mainly used in hybrid renewable energy storage media to support other storage media. Lithium-ion (Li-ion), sodium sulfur (NaS) and related zebra battery (Na-NiC12), nickel cadmium (NiCd) and related nickel metal hydride (Ni-MeH), and lead acid (Pb-acid) batteries are flow batteries, among which Li-ion shows the best performance. A Li-ion storage capacity of 100 MWh (almost equal to that of a wind farm) is theoretically possible to be installed by chaining the batteries, which gives a new dimension for renewable energy use [HAL 08].

Superconducting magnetic energy storage (SMES) has almost zero energy loss, which achieves 98% efficiency. This technology has a lifetime capacity of 1–10 MW, which is mainly used in Europe.

In flywheels, kinetic energy is stored by rotating cylinders. Although the lifetime capacity of flywheels is longer than 20 years, they are very expensive and occupy a large area.

Thus, research in the energy storage field is focused on making it cheaper and its lifecycle longer with planned charge and discharge operations.

Despite the continuous progress in energy storage, we are still not able to store heat from sunlight during summer for use in winter.

2.2. Technology for energy distribution

Energy distribution technologies include all the machines and utilities used for transmission and distribution as well as distribution reliability and resiliency to ensure energy transfer security.

An adaptive voltage control system is the most critical unit in distribution. The core determines tap change requirements, selects controlling voltages and the capacitors to be used. Most of the research conducted has shown no damage to equipment or overheating of motors until much lower voltages are reached and sustained. The core runs on feeder monitoring devices that monitor the feeder kWh, kVARh, kW, kVAR current and voltage of the substation data concentrator and controller (SDCC). The power monitoring devices accommodate different manufacturers of regulators, load tap changers. The operator interface allows an operator to determine the status of the system, turn it on and off, and, in some cases, to manually operate regulators and capacitors. There can be multiple line voltage monitors on a single feeder. Line regulator and capacitor control interface units (LRCCI) are microprocessor-based units located at mid-line regulators and switches. In currently operational AdaptiVoltTM systems, both the SDCC and core units are PLC-based and they communicate with each other. The system provides distribution system visibility to supervisory command and data acquisition (SCADA) systems as well as information that can be used for engineering, planning and operational use [WIL 04].

SCADA systems are an important strategic asset of a power distribution business, which are used for remote controlling. Sequence of events is recorded through automated digital data collection. SCADA systems typically implement a distributed database, referred to as a *tag database*, which contains data elements called *tags* or *points*. A point is a value-time pair. It represents a single input or output value if it is hard or represents a tracing data for the system, following an operation and is called the *soft point*. Switch, alarm information, pressure and capacity are typical points.

With the growing number of small grids, automation of power distribution has become more important. The following fields are considered as the main research subject in the regional distribution:

– self-healing;

– high reliability and power quality;

– resistant to cyber attacks;

– control of a wide variety of distributed generation;

– storage options;

– optimization of generated power utilization;

– minimization of operations and maintenance expenses.

Research has shown that all the energy-efficiency efforts make energy sales reduce, although the number of energy sales companies increases with increasing liberalization of utility companies. The load and demand data monitored through mobile devices will give the lead for sales.

2.3. Technology for improving energy use

Energy saving is the major goal of technological developments in energy use. At the end of the 20th Century, Weber predicted the improvement potential for energy use as 30% [WEB 97]. The first improvements were made for using steam in heating with a higher efficiency. Changes in steam engine technologies have contributed to the productivity development of manufacturing sites. However, technological innovations in electrification have an important impact on heating as well as power utilization by the residential sites as well as the industry. New techniques have been widely used in information and computer technologies for electrification and for providing power and heating in a district, where district heating has become the subject for the residential energy [REZ 12]. Russia, where 40% of energy was spent on heating, has shown progressive efforts on technological innovations. Some of these technologies can be summarized as efficient burners, automated combustion control systems for boilers, polymer insulation of pipelines, bioreactors and biogas collectors, multi-level heat supply regulation, automated water filters, thermal insulation, new valves and compensators, heat reflectors, low e-films on windows and so on [GRI 97].

The influence of energy consumption on climate change has motivated the renewal of energy investment and management strategies globally. The Intergovernmental Panel on Climate Change (IPCC) has suggested that global warming necessitates the collaborative international research to avoid extreme weather conditions, ocean floods, loss of biodiversity, poor harvest and arising variations of illnesses [WHI 90]. The IEA has presented new strategies in World Outlook reports from 2000 onwards, suggesting developed countries adopt new energy resources and energy-efficiency strategies for global sustainability [INT 02].

2.3.1. *Energy-saving technologies*

Energy consumption quota, standards for pollution and environmental taxes are the policies that elicit the biggest energy consuming industries to invest in energy saving. The technologies used for less energy consumption without reducing the productivity, alters with respect to the products produced. For example, in the iron and steel industry, pulverized coal injection technology, continuous casting technology and scrap steelmaking are widely used by big firms instead of basic oxygen furnaces [ZHA 08]. Dielectric heating or high intensity ultrasound are considered novel energy-saving techniques in the food industry [PER 10]. In the automotive industry, the focus is on reducing the amount of energy used by assembly equipment, joining technologies, heating equipment and painting processes as well as industrial robots [BOR 16]. There are commonly used technologies like electrical motors that consume the majority of the energy used in these industries. Electrical motors are responsible for 75% of power consumption during manufacturing processes [SAI 10]. Significant savings can be realized by improving air volume in air conditioning, water pumping and exhaust air systems, where the majority of patents in the field are approved.

Figure 2.7 shows the amount of electricity used by different motors. Computer numerical control (CNC) machining is another type of machine used in multiple industries. Optimization of cutting is the most energy-saving activity in CNC machines. Manufacturing control strategies can be redesigned to include ecological concerns in addition to the economic objectives.

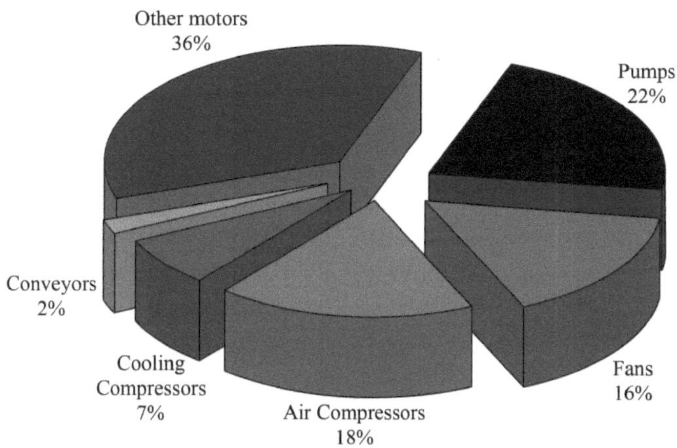

Figure 2.7. *Power utilization share of electricity motors [SAI 10]*

All these efforts trigger the requirements for measuring the net amount saved. Energy-saving technologies are mainly based on the forecasts and optimization of energy use.

2.3.2. *Efficient energy use in manufacturing processes*

Energy efficiency is measured as the ratio of the productive output of the manufacturing system within the pre-defined quality to the total energy consumed by the processes.

In the production processes, this definition allows us to focus on minimizing the energy requirement, decreasing energy consumption in operations or using efficient start-up and/or shutdown processes. In particular, efficient machine components will be enough to improve the efficiency of using alternative energy. In the entire manufacturing plant, minimizing waste and lean approaches are applied as well as employing the shifts of time and location. Inbound logistics are handled in lean manufacturing approaches because the transportation in the plant is not a value adding activity. However, industrial buildings need smart approaches to optimize the heating and cooling conditions.

The European Union defined mandatory rules and targets for energy efficiency in manufacturing. European manufacturing companies still consider greening as green product manufacturing, whereas new processes demand more energy than the energy saved by the green product. It is obvious that if companies aim to realize process innovation, they are more inclined to improve energy efficiency, whereas product innovation does not directly result in efficiency improvements [GER 16].

The US Government has always been supportive of any technological development in the energy field to reduce foreign dependency. Following the 1970 crisis, governmental support was focused on security, sustainability and price reduction. Huge support was given to nuclear fission, fusion energy and renewable energy. Towards 2000, initial cuts started on nuclear fission and fossil research. After 2000, research focused on the wind turbines, solar photovoltaics, shale gas and turbines. A symposium on 21st-Century manufacturing discussed the importance of combining renewable energy investments and the manufacturing research for innovation [WES 11]. The collaboration of the USA and China on the innovation for the wider use of renewable energy technologies was also discussed in the same symposium as China and sets an example for other energy agencies while they are converting from coal to wind power.

The hardware and software aspects of energy consumption in the mobile platform is one of the new considerations due to increasing power demands. Initially, studies were focused on the hardware products only, but the lucrative evolution of green products supported the software companies to take greater initiatives [KER 15]. Figure 2.8 shows the ubiquitous access of energy management including energy load, control and measures followed by mobile and Web access.

Research on cost-effective energy savings in buildings and traffic is continuing. Many advanced technologies, including LED lights, innovative window systems, new types of integrated cooling systems and power-saving electronic devices, are commercially available already. In addition, new commercial buildings with low overall energy use are presented to the market.

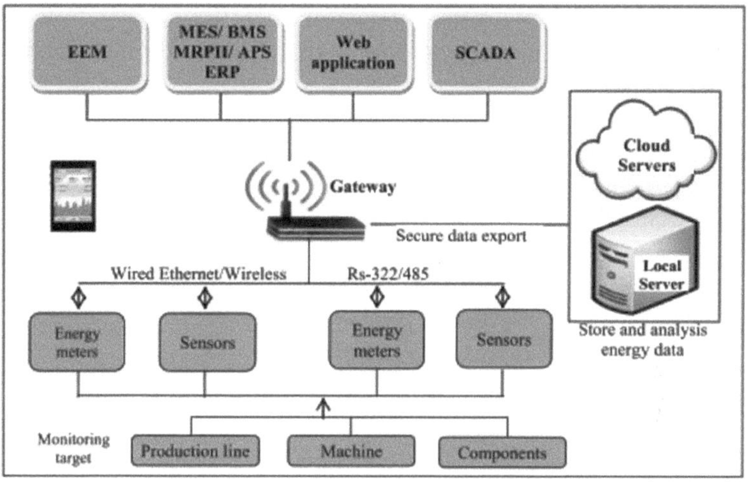

Figure 2.8. *Energy use monitoring using the Internet [SHR 15]*

Innovations in the field of energy were initially triggered by cost and economic reasons, but the technological push has now added to this. It is observed that those who invest in electronic and optical products give prominence to research and even obtain patents on energy technologies. Renewable energies are more significant in the chemical product manufacturing industry. Computer and telecommunication industries give importance to energy conservation. Manufacturers of motor vehicles and/or machinery invest almost half of the electrical equipment in energy conservation [REX 15].

2.4. Optimization techniques for energy

Optimization of energy use begins with resources: the first challenge involved creating heat with the least amount of wood.

Finding the wood to burn was not easy without the right cutting tools. The collection of wood pieces caused time and material expenses for gathering and sheltering, respectively. Newton's gravitational theory provided the first quadratic optimization without constraints, and thus allowed for heating optimization [NEW 87]. Lagrange multipliers are defined by the facilitation of equilibrium definitions in the first published work of Lagrange in 1854 [LAG 15]. Maxwell initiated the optimal control theory through steam feedback control in 1868, although it was defined by Pontryagin 50 years later as a great step in optimization [ROS 09]. The paradox of energy efficiency and manpower is faced when coal is used as the main resource of heating. The Jevons paradox was introduced in economics based on *The Coal Question* [JEV 66]. It is interesting to discover Jevons as a follower of Newton, who developed the generally accepted famous method for finding an unconstrained minimum of a twice continuously differentiable function.

THE

COAL QUESTION;

AN INQUIRY
CONCERNING THE PROGRESS OF THE NATION,
AND THE
PROBABLE EXHAUSTION OF OUR COAL-MINES.

W. STANLEY JEVONS, M.A.

SECOND EDITION, REVISED.

London:
MACMILLAN AND CO.
1866.

Figure 2.9. *The origin of the Jevons paradox: book cover of* The Coal Question *[JEV 66]*

The Jevons paradox is the most well-known paradox in environmental economics because it occurs when technological progress increases the efficiency with which a resource is used (coal in his case), but the rate of consumption of the same resource rises because of increasing demand. The paradox is only solved by Newton's unconstrained quadratic minimization problem in each step. Starting from an initial point, it generates the sequence, which under certain conditions can be shown to converge to the minimum of the original function. To treat the case, several modified Newton methods based on quadratic models different from the second-order Taylor series approximation have been developed. Also, the mathematical theory of quasi-Newton methods for unconstrained minimization has been developed through the study of quadratic models [DEN 83, BAZ 93].

The history of optimization and energy optimization is extensive. Energy is a field that has always triggered new improvements in optimization, and hence optimization is applied at any small field and/or function in energy production and management.

Ash and charred bone, the earliest known evidence of controlled use of fire, reveal that human ancestors would have used fire a million years ago, a discovery that researchers say will shed light on this major turning point in human evolution; for example, the archaeologist Michael Chazan stated that "socializing around a campfire might actually be an essential aspect of what makes us human. The control of fire would have been a major turning point in human evolution" [CHA 12].

Economy and optimization interact with an impact on both sides; that is why the optimization is widely used for minimization of costs and maximization of profits. Optimization affects technology, culture and environment. This means that technology for energy is used to optimize the savings. In addition, the environmental effects must be optimized so that ecological life is conserved. Moreover, social culture should be widely opened for all the new technologies, which is only possible when knowledge is optimized.

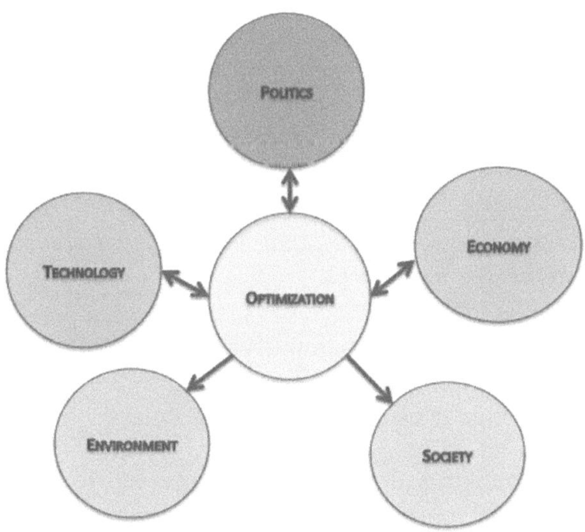

Figure 2.10. *Optimization impact*

Several million years ago, coal and oil formed to become the first energy resources. Nuclear energy was only declared as a clean energy resource in the 20th Century. Despite this forests are being still demolished to generate heat according to the two thermodynamic rules below.

2.4.1. *Thermodynamic rules*

The two following thermodynamic rules, which are the basic outcome of fire, make energy a complex concept that cannot be accepted as merely another input [ERN 78].

The first law of thermodynamics is the efficiency measure where transfer is defined as:

$$\varepsilon_1 = \frac{\textit{energy transfer by a device or system}}{\textit{energy input to the device or system}} \qquad [2.1]$$

This could be easily explained by an example from a power generator unit: a given amount of Btus of high temperature steam in the water circulating unit has greater capacity to do work than the same amount of steam with a lower temperature. The quantity of energy is the same in both cases, but the transformation of lower temperature involves a reduction in the quality of Btus and loss of work capacity. In other words, useful energy consumed is measured by the changes in the ability to perform useful tasks.

The second law of thermodynamics shows the availability expressed as:

$$\varepsilon_2 = \frac{\textit{Minimum work required to perform a given task}}{\begin{array}{l}\textit{Maximum possible work that couldbe be extracted}\\ \textit{from the fuel consumed}\end{array}} \qquad [2.2]$$

For example, a house heating pump usually has an efficiency of 0.07, but with technology, this improved the efficiency up to 2.5 times. If there is too much improvement, then it makes the first law useless.

For a heat exchanging device, it is important to minimize thermal irreversibility, but it is more important to minimize the mechanical power lost in achieving a prescribed heat exchange performance: for this purpose, a relative irreversibility index called the Bejan number is used to assess the merit of a given configuration [SCI 96].

A key point is heat loss minimization during heat transfer, which is studied to determine whether energy resource in any kind of heat generation including the storage is fossil or renewable.

Optimization of energy use is used for the functions of investment, design, exergy, operations and performance [ERC 15]. These functions are repeated at any phase of the supply chain, for energy generation, distributions or sales, and mainly for energy use.

Thermodynamic laws could be considered as the first instances of Industrial Optimization [THO 54]. Cost Minimization and the design were handled together with an observation of the energy production site as a whole [ZHU 00].

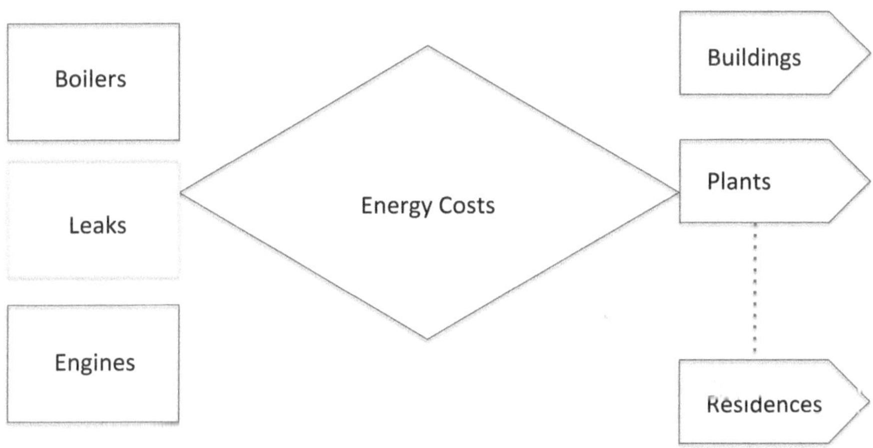

Figure 2.11. *Simplest optimization model for heating*

In the majority of the optimization studies on heating, the problem is based on the analysis of primary energy costs, optimization of the energy infrastructure, optimization of the operational costs for energy generation and the supply and demand equilibrium. Demand of energy for heating varies with time intervals.

2.4.2. *Investment*

The main goal of the investment problems is minimizing the total cost with a simple linear or nonlinear model used [DAG 00, THO 09]:

$$\begin{aligned} &Min\ Z = f(x,y) \\ &s.t. \\ &\quad g(x,y) = 0 \\ &\quad h(x,y) \leq 0 \\ &\quad x \geq 0, y \in \{0,1\} \end{aligned} \quad [2.3]$$

where an objective function f(x,y) like a cost function is minimized. The objective is to satisfy under equality constraints as g(x,y) = 0 like power demand and supply equations, or inequality constraints h(x,y) ≤ 0 describing the limits in capacity for the resources or technologies in the system. The variable x is continuous and corresponds to decision variables like energy and material flows, while the variable y represents the binary variables that are used to determine the existence or use of certain resources and/or technologies.

The most common investment problems are the allocation with minimum investment cost as described in Perrin and Sibley [PER 93], or minimum investment cost mainly for power generation as put forward in an early example [AWE 93]. Two kinds of models are observed, the static and dynamic models. The static ones that keep the environmental criteria, like temperature stable, assume relative cost certainty and low rates of technological progress, while homogeneity is observed in technology, with infinite time. Today, it is impossible to analyze without the consideration of dynamic changes.

Investment optimization models, in general, use the following Euler–Lotka equation [LIK 03]:

$$1 = \int_{a_p}^{\infty} R(a)P\left[a > a_p\right]\exp\left(-ra\right)da \quad [2.4]$$

where R is the fitness to the state that is maximized with a probability of being in state *a*. A majority of applications take the maximum fitness *r* as a constant like 1 and benefit the Taylor equation to make it a simple differential equation as in equation [2.5]:

$$1 = \int_{0}^{\infty} e^{rt}\ell\left(t\right)b\left(t\right)dt \quad [2.5]$$

However, today's energy planners should have a broad diverse range of technological and institutional options for energy generation and plan for a future that is dynamic, complex and uncertain. Clearly, more powerful techniques are required for robust strategies that remain economical under a variety of uncertain future outcomes. Financial investors are learning to deal with uncertainty, and a portfolio of assets provides the best means of hedging future risks [AWE 06]. Expected portfolio return $E(r_p)$ is the weighted sum of individual expected returns, where X_1 and X_2 are the shares of assets in the portfolio:

$$E(r_p)=X_1E(r_1)+X_2E(r_2) \qquad [2.6]$$

In contrast, the portfolio risk σ_p is expressed in two securities with the correlation ρ_{12}:

$$\sigma_p=\sqrt{X_1^2\sigma^2{}_1+X_2^2\sigma^2{}_2+2X_1X_2\rho_{12}\sigma_1\sigma_2} \qquad [2.7]$$

An example for profit maximization in energy R&D is modeled by Shittu and Baker, introducing tax on carbon emissions. They use a two-period theoretical model. Investments in R&D are made in the first period in anticipation of a future carbon tax. For simplicity, the production in the first period is ignored. Optimal production is chosen in the second period after the firm learns about the carbon tax and technical change is achieved, leading to a second-period profit function [SHI 10].

In some of the latest works, focus is made on multiple energy source investments with multiple objectives [KAY 15]:

$$\begin{aligned}&\mathit{Maximize}\ \lambda\\&\text{s.t.}\\&\qquad \lambda\le\mu_g(x)\\&\qquad C_h.x\le b_h \quad \mathit{Maximize}\ \ \lambda\\&\qquad x\ge 0\end{aligned} \qquad [2.8]$$

where the complex relation $\lambda \le \mu_g(\vec{x})$ makes it difficult to solve.

To use the well-known differentiable optimization techniques, different approaches are chosen. For example, to minimize the risk of a portfolio of renewable energies, different renewable energy projects are evaluated. To validate the assessment, economic models are developed. After evaluating the projects economically, the risk of all renewable energy projects is minimized. Another example is the use stochastic models. The portfolio is composed of a certain percentage of each technology to make up the whole 100%. This portfolio has an associated return and risk that depends on the percentages of each technology. The internal rate of return (IRR) changes with the risk defined [MUÑ 09].

Some researchers just linearize everything as in the distribution of electricity, water, gas and heat [PTA 06].

Investment optimization will continue to be used with the interest of different investors (energy generator, city government, energy distributor and energy consumer).

2.4.3. *Architectural optimization*

Architectural optimization is structured to find the best utilization of utilities, materials and the technology. In early research on solar cells, Omer constructed an optimization model for the use of a thermoelectric device that was developed for geometrical optimization of thermoelectric element legs and prediction of the performance of an optimum device in power generation mode [OME 98]. This objective of design optimization has continued until now, and will continue in the future with the improvements in urban or district energy management [BAL 15a, BAL 15b].

Thompson defines the design optimization for the environment in the simplest way [THO 99]. He includes the objectives of efficient use of resources for energy generation, service life, reuse of wastes and minimizing the variety of materials.

He indicates that he was inspired by Leonardo Da Vinci's 500-year-old observation on the limitations of human creativity relative to the creation of the natural world, which still holds.

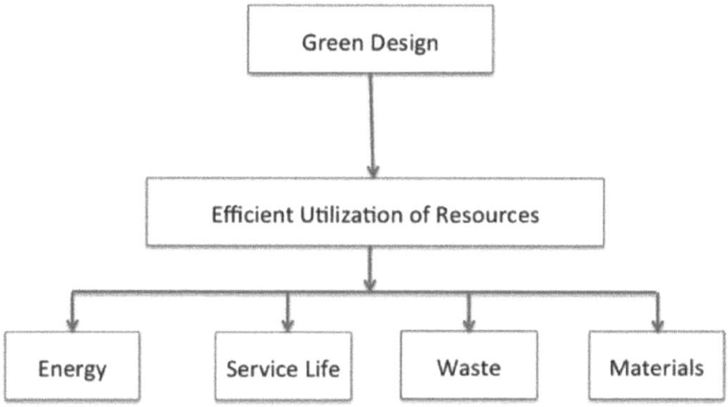

Figure 2.12. *The primary attributes of design for the environment [THO 99]*

Design optimization is largely considered during the technology choice or network designs [LUI 03]. Lots of application studies combine mathematical optimization models with financial models for renewable energy generation sites as in wind energy farms [KIR 01]. In industry, design optimization is crucial for generating energy to respond to the requirements of the production site itself. Today it is very important even in micro-cogeneration [JIN 12a, JIN 12b]. Waste recovery optimization is another industrial concern that is in need of design optimization [PIE 13].

Most of the design optimization studies have dealt with investment optimization and/or maximization of exergetic efficiency. That is why design optimization is an inevitable part of exergo-economic or thermo-economic studies mostly used for heating systems [TSA 93, CAS 04].

2.4.4. *Exergy optimization*

Exergy is defined as the maximum amount of work that can be produced by a system when it comes to equilibrium with a reference environment. Exergy analysis has proved to be a powerful tool in the thermodynamic analyses of energy systems. It is used to quantitatively evaluate the causes of thermodynamic imperfection of the process under consideration. Exergy analysis usually aims to determine the maximum performance of the system and identify the equipment in which exergy loss occurs, and indicates the possibilities of thermodynamic improvement of the process under consideration. Exergy analysis of a complex system is analyzed for each component of the system separately. Identifying the equipment in which the main exergy loss occurs shows the direction for potential improvements. The exergy efficiency for a waste heat recovery system in the cement plant can be expressed by:

$$\eta_{ex} = \frac{E_{in} - \sum_i I_i}{E_{in}} \qquad [2.9]$$

where E_{in} is the inlet exergy and $\sum_i I_i$ is the sum of the exergy losses of processes. For the co-generation system in the cement industry, the exergy input comes from the preheater exhaust and clinker cooler exhaust gases, and the exergy losses occur in main components, such as AQC boiler, SP boiler, turbine, pumps, condensers, heat exchangers and flasher [WAN 09]:

$$E_x = R.T.\sum_i P_i \ln \frac{P_i}{P_{i,eq}} - \left(P_i - P_{i,eq}\right) \qquad [2.10]$$

Thermo-economics, on the other hand, combines exergy analysis and economic principles. Maximizing the exergy is the main purpose of exergy optimization studies [NIE 95] where exergy function is simply given as in equation [2.9].

In the exergy equation, R is the gas constant (8.317 J/(mol.K)), T is the absolute temperature in Kelvin, Pi is the concentration of matter in the compartment i and $P_{i,eq}$ is the concentration of the compartment i.

Exergy optimization is mainly used for the process optimization and lead to exergo-economic studies by considering the energy conversion, energy use and energy saving in the same model [ZHA 00].

Based on the given boundary conditions of E_{xOi} and E_{xDi}, which are the outcome of the optimization of the other two links, we can optimize the energy conversion subsystem to obtain the highest exergy conversion efficiency or the minimum cost of the effective-supply exergy E.

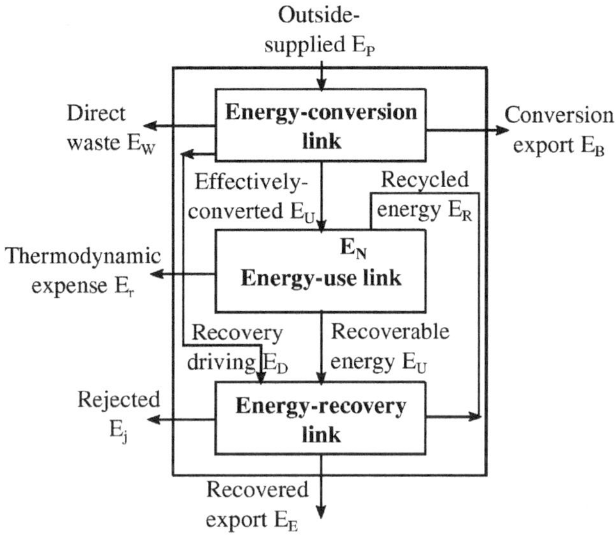

Figure 2.13. *Three link model [ZHA 00]*

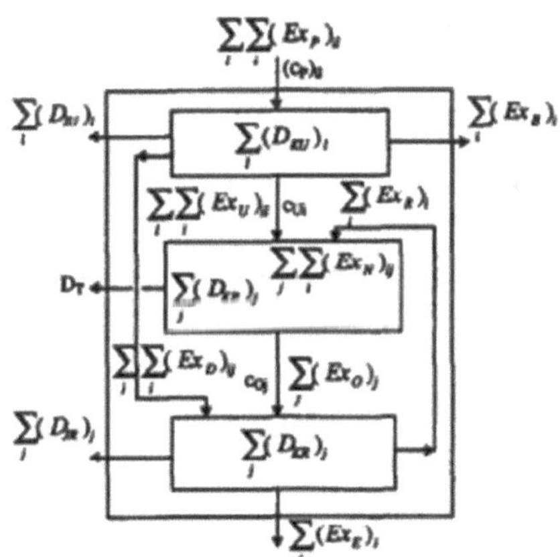

Figure 2.14. *Hierarchical exergy model [ZHA 00]*

The objective function of the subsystem has been driven from the money balance equation of the energy conversion link:

$$Min\, C_U = \left(\sum_i C_{pi} E_{xpi} - \sum_i C_{Bi} E_{xBi} + \sum_j Z_{Uj} \right) \times / \sum_i \left(E_{xUi} + E_{xDi} \right) \qquad [2.11]$$

Optimization of this subsystem involves combustion gas turbine, furnace co-generator, back-pressure steam turbine, boilers and other effective measures to improve the exergy conversion efficiency.

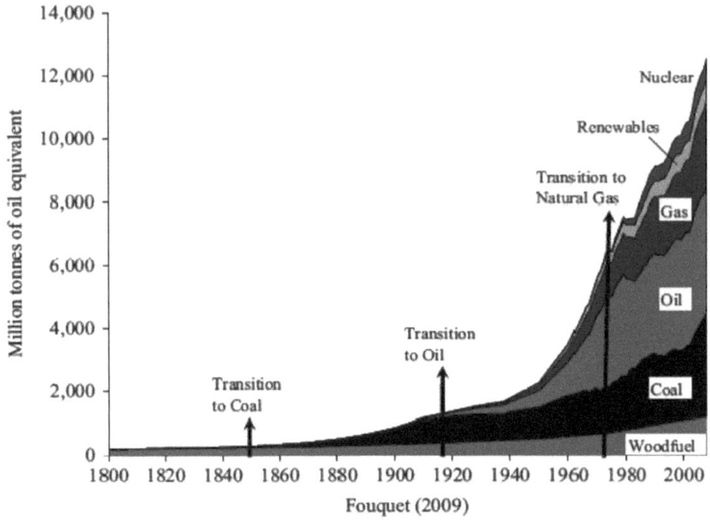

Figure 2.15. *Transitions in energy use [FOU 12]*

If the site is non-operational or used, then the amount of steam demand is null. However, if the site is operational, the demand may change by the day or the hour if detail is requested. This shows a nonlinear (non-deterministic) curve for the heat demand. Researchers suggest that the strong nonlinearity in heating demand explains the necessity for heat sources to be distinguished [WOL 98].

Initially, the heating source was wood, which turned to be coal when lignite was discovered. Over the years, there has been a transition from coal to oil, to natural gas and to renewables and nuclear energy, as shown in Figure 2.15. Any heating demand is dependent on oil prices, and even nonlinearity cannot answer for the dependency on the oil price.

2.4.5. *Optimization for industrial energy saving*

Towards the end of the 19th Century, it was recognized that equilibrium conditions in the electrical network were attained when the total energy loss was minimized [GAS 01]. It was proved that the sum of energy losses in the equipment used for the electrical network is a quadratic function of the branch currents when all devices are of a linear nature [DEN 59]. Dennis has also shown that the Kirchhoff equation can be used in the equality problem by adding the Lagrange multipliers for each network node. His model has ever since been used to solve the problem of power flow through the transmission network in order to meet a certain demand.

Energy should be provided in function of demand. Economists handled this equality problem using the strategies constructed by the application of game theory. It aimed to give solutions for conflicting issues in the balance. Dantzig, Gale, Kuhn and Tucker contributed to this goal by using mathematical modeling.

The equations of the models are as follows [BAS 82]:

– Objective function:

$$\sum_{t=1}^{T} \beta(t)pr(t)+cx(t)+ky(t) \qquad [2.12]$$

where t is the current index of time period, T is the numbered time periods, $\beta(t)$ p is the vector of energy resources prices, r is the resource activities, c is the current cost vector, x is the vector of energy conversion activities (LP variables), k is the vector of capital (investment) costs and y is the vector of capacity increments (LP variables).

– Resource availability:

$$\sum_{j=1}^{J}\sum_{t=1}^{T} r_j(t) \leq K_j \qquad [2.13]$$

where r(t) is the annual extraction in period t and Kj is the availability of the resource j.

This constraint is defined for each resource and category.

– Resource requirement:

$$\sum_{j=1}^{J}\sum_{t=1}^{T} r_j(t) \geq \sum_{t=1}^{T}\sum_{l}^{L}(\upsilon_l x_l + w_l y_l) \qquad [2.14]$$

where j is the index of resource category, J is the number of resource categories, υ_l is the specific consumption by production activity x_l and w_l is the inventory requirement for capacity increment y_l

These equations are defined for each time period and resource kind.

– Capacity equations:

$$x_j \leq (\sum_{\tau}^{t} y^{\tau}) h_j f \quad \forall j \text{ and } \forall t \qquad [2.15]$$

where j is the index of load resource , h_j is the load duration of the resource j and f is the plant factor. This equation is defined for each time period and technology.

– Demand constraints:

$$\sum_{s}^{S}\sum_{j}^{J} \eta_{sj} x_j \geq D_s \quad \forall t \qquad [2.16]$$

where s is the index of demand sector, η_{sj} is the conversion efficiency (or equal to 0 if x does not supply j to demand sector s) and D is the annual demand.

This equation is defined for each time period, demand sector and load region.

– Build-up constraints:

$$y^t \leq Gy^{t-1} + g \quad \forall t \qquad [2.17]$$

where G is the growth parameter and g is a constant, allowing for start-up.

This equation is defined for some (primarily new) technologies and for each time period.

– Emissions constraints:

$$\beta \sum_{i}^{I} em_i x_i \leq Z\% \quad \forall t \qquad [2.18]$$

where β are total emissions (LP variables) and em_i are specific emissions of technology i. This equation is defined for each time period.

It is often difficult to apply the fact that energy types are to some extent substitutable in end-use demands [ERN 78]. This difficulty leads us to stochastic models. Hybrid use of renewable energies is presumed to be the trend for coming decades.

2.4.6. *Operational optimization*

Operational optimization is a critical task for energy generation units. There is a balance to embrace, as excess energy is wasted and shortcomings in response to demand leads to low consumer satisfaction. Periodic operation plans or schedules have been the output of optimization studies. The supplied power and the demand have been considered to minimize the costs under the constraints of restricted capacities of energy generation units. Uncertainties in measuring renewable resources, such as hydro-power or wind power lead the operational optimization studies. The pioneers study the forecasted resources in line with capabilities of power production [CAS 04]. For the energy load of combining hydro-energy, wind power and reservoirs, the following is a good example, with P_{wi} being the power produced by wind in period i, P_{hi} the power produced by hydro in period i, E_1 the initial reservoir level, E_M the reservoir capacity with a change ΔE_M responding to the demand D_i in period i and generation efficiencies η_w, η_h:

$$\begin{aligned}
&Max\,Power = \sum_i (c_i(P_{wi}+P_{hi}) - c_c P_{wci}) - c_M \Delta E_M - c_{E1}\Delta E_1 \\
&s.t. \\
&E_{i+1} = E_i + t(\eta_w P_{wi} - \eta_h P_{hi}) \\
&P_{wi} + P_{hi} \geq D_i \\
&E_i \leq E_M \\
&P_{wi}, P_{hi}, E_i \geq 0
\end{aligned} \qquad [2.19]$$

Lifecycle costing or analysis has been a comprehensible analysis in operational costing. For example, operation costs for the grid are calculated using equation [2.20] [MOR 14]:

$$Max \quad NPV = \sum_{j=1}^{N} \frac{d}{1+p} (\sum_{1}^{8760} (D_{ij} C_{ij} - ED_{ij} + K_{ij}) - OM_j + PR_j) - I \qquad [2.20]$$

where NPV stands for net present value of N projected years at year j. Power demand (D_{ij}) at electricity cost (C_{ij}), environmental dispatch (ED_{ij}), and reliability obtained by micro-grid benefits (K_{ij}) are calculated on an hourly basis, but the annual operation and maintenance cost (OM_j) and profit obtained with improved reliability (PR_j) are considered annually. The initial investment I is carried out with a price rate p.

Uncertainty is caused by resource availability or seasonal energy demands which makes it difficult to operate the turbines for generation, and storage for limited capacity has to be planned to minimize costs [PAP 98]. Today, operational optimization still minimizes the cost of operation, which is extended to include demand site management, storage optimization and even real-time forecasting [WAN 15].

2.4.7. *Performance optimization*

Performance optimization for energy systems involves the interaction between growth rates and energy supply costs. The most well-known examples are the decline in productivity growth rates in 1973–74 and the OPEC oil price shock in 1979 [BER 90]. Another area of major performance studies is the lightning or heating performances of building designs. Industrial Energy generation has influenced both energy performance and production performances, hence optimization is effectively used in that field.

One of the typical applications of performance optimization is Primary Energy Consumption (PEC), which is defined as the amount of energy consumption, plus losses that occur in the generation, transmission and distribution of energy [CHO 09]. The objective function to minimize PEC can be expressed as:

$$Minimize\sum_{t=1}^{T}\{c_{el}(t)E_{grid}(t)+c_{f-ngu}(t)F_{pgu}(t)+c_{f-boiler}(t)F_{boiler}(t)-c_{el-ex}(t)E_{excess}(t)\} \quad [2.21]$$

This is an example where the grid, boiler and excess energy are taken into account, each of which has costs as a function of time.

More enhanced studies of performance optimization are run on clusters of potential energy resources [LAM 10]. In this example, the objective is to minimize the total carbon footprint (CFP) within the region defined as a function of footprints of each zone (CFP), fuel consumption (FC), carried distance (Dist), biomass load (B), lorry capacity (C) and carbon emission factor (CEF). Two constraints are taken into account, namely biomass carried to other zones cannot exceed the biomass produced and bioenergy delivered to the zone cannot exceed the deficit in that zone.

$$\begin{aligned}
&MinimizeCFP=\sum_{ij}CFP_{ij}\\
&s.t.\\
&CFP_{ij}=FC_{ij}\times Dist_{ij}\times\frac{B_{ij}}{C}\times CEF\\
&\sum_{j}B_{ij}\le A\times B_{i} \qquad \forall i., \;\; B_{ij}\|_{i=j}=0\\
&\sum_{i}HV_{i}\times B_{ij}\le D_{j} \quad \forall j\\
&B_{ij}\ge 0
\end{aligned} \qquad [2.22]$$

Industrial performance in energy industry is another subject of research, which is measured through the performance of energy and exergy processes [PTA 06].

2.4.8. *Leak optimization*

Heat leak is to minimized when there are finite size capacities [BEJ 96]. Q_{opt} is the heat leak for cold temperature, the coldest temperature of the system T_L, to the room temperature T_H with conductivity $k(T)$ trying to reduce the temperature; and related to the ratio of the cross-section of the conductor and the length of the conductor . The enthalpy S_{gen} should also be minimized in parallel. The objectives to be optimized can be expressed as follows:

$$Q_{opt} = \left(\frac{A}{L} \int_{T_L}^{T_H} \frac{k^{1/2}}{T} dT \right) k^{1/2} T \qquad [2.23]$$

$$S_{gen} = \frac{A}{L} \left(\int_{T_L}^{T_H} \frac{k^{1/2}}{T} dT \right)^2 \qquad [2.24]$$

Industrial use or efficiency needs to consider the interactions with economic factors. To express the problem for optimization, either econometric models or mathematical models are constructed.

2.4.9. *Strategic optimization*

Industrial companies that have to face a strong competition need individual policies. In macroeconomic approaches of country-wide strategies to market balancing, regulations or policies for investment in different energy resources are generally handled using the game theory approach. An individual industrial company can also use the game theory approach for defining competitive or efficiency strategies.

Game theory was defined as a card game by Waldgrave in 1713, but defined as a scientific study of multiplayer decisions after the Tennis league evaluation in 1997 [WIR 05]. During a game, the players choose simultaneous actions and receive payoffs that depend on the combination of those actions. If the payoffs are common knowledge among all players, then it is a complete information case for a static (simultaneous move) game [GIB 96]. The normal-form representation of a game defines:

– the players;

– the strategies;

– the payoffs for each strategy.

Individual strategies of the players have an impact on each other. To allow each player to choose the strategy freely, the classical approach is to go for the balanced games defined by Nash equilibria [NAS 50], where the equilibria is to be constructed. Cournot's duopoly model, defined in 1882, has been widely accepted in economics and assumes that equilibria exist. The game can be constructed as a cooperative game where the common benefits are considered, or as an uncooperative one where each player chooses his strategy with the aim of maximizing his own benefits without caring about others. The game is defined with the Bayesian approach when deterministic information does not exist.

We will discuss the game theory approach by considering energy supplier strategies in a competitive position.

In a liberated retail energy market, each electricity supply contract is signed with the consumer for a certain price. Jia and Yokoyama [JIA 03] define the following game for the suppliers to profit.

The generation cost for the generator i is represented by $f(i)$ and per hour profit for generator i is represented by $\nu(i)$. The functions can then be defined as follows:

$$f(i) = a_i P_i^2 + b_i P_i + c_i$$
$$\nu(i) = \rho_i P_i - f(i) P_i - \omega P_i \quad [2.25]$$

where ρ_i is the price of the generator i given to the customer (€/MWh), P_i is the schedule effective power of the supplier i (MW/h), ω is the transportation charge rate (€/MWh) and a_i, b_i and c_i are the coefficients for the generator i.

Facing the same customer, if the generators cooperate with other generators, then they have to obey the equal price. If for a single generator, the derivative of the cost function is:

$$\frac{df(i)P_i}{dP_i} = 2a_i P_i + b_i = \lambda$$

and power is:

$$P_i = \frac{\lambda - b_i}{2a_i}$$

total power becomes

$$P_1 + P_2 + ...+ P_n = P_R$$

$$P_R = \frac{\lambda}{2}\sum_{1=1}^{n}\frac{1}{c_i} - \frac{\lambda}{2}\sum_{1=1}^{n}\frac{b_i}{c_i} \qquad [2.26]$$

hence

$$p_i = \frac{1}{2c_i}\frac{\sum_{i=1}^{n}(b_i / c_i) + 2P_R}{\sum_{i=1}^{n}(1/c_i)}$$

Therefore, in a coalition S whose number of generators is m, profit is:

$$v(S) = \sum_{i=1}^{m}(\rho_i P_i) - f(i)p_i - \sum_{i=1}\omega p_i$$

where λ is the Lagrange multiplier, P_R is the total load (MW) and p_i is the effective power generated after cooperation (MW/h).

The Shapley value is used to calculate the expected marginal amount of each energy generator to the coalition in a particular concept of fairness that can be achieved by using equation [2.26]:

$$\phi_i = \frac{(m-1)!(n-m)!}{n!}\{v(S) - v(S-i)\} \qquad [2.27]$$

where only *m* of the whole *N* energy generators are in the coalition of *S* and *S-i* is the case where the ith generator is not included in the coalition.

As the source of energies increases, investment decisions are made using the game theory approach as proposed in [RAI 10]. In that paper, the game is constructed on energy resources of diesel, natural gas, coal and hydropower to find the annual investment costs and the average marginal costs based on the peak load prices.

With the increase in district heating and the regional or local grids, game theory is widely used in balanced use of energy in demand-site management. Lo Prete [LO 16] gives a good example of cooperative games that are constructed in the micro-grids with multiple objectives considering the fuel costs, heat rates, operation costs, CO_2 emissions, generator lifetime and ramps or peaks.

2.4.10. *Uncertainty handling*

Uncertainty is widely observed in the field of energy. Discrete optimization models are used as long as the limits and constants of a model are deterministic. However, when short-, medium- and long-term energy requirements and supplies are studied, the first uncertainty is caused by the demand. Energy loading depends on demand, hence it should be calculated to estimate the values of demand. The second uncertainty is the price. Electricity or natural gas prices change by markets like the European market or North American market. Furthermore, since the price is the result of a market balance between energy supplied and energy demanded, it changes by hour. In liberated markets, the retail market works on the day-ahead prices as well as the intra-day prices. The major uncertainty is caused by the risks in the market.

Stochastic optimization approaches are designed to handle uncertainties. Wallace and Fleten [WAL 03] give a review of stochastic models applied in the energy field. They claim that these methods have been used in energy scheduling since 1947. For example, if the demand is observed for a whole year, there are base load and peak load hours. However, when it is defined as a discrete problem, we obtain a quadratic objective function that causes some losses. If the probability of base and peak hours is observed, then a stochastic model can be constructed. In [VEH 05], there is a good example of calculating the power prices in the spot market using the stochastic factors where demand is estimated according to the temperature.

The cost of an energy investment project, which is already scheduled by considering the known k number of risks for which risk reduction actions can be taken, is constructed in a predictive stochastic model, as described in [ZAF 07]. The cost of the project until a certain time instant *t* is considered as follows:

$$y(t) = y(t-1) + T(t) + R(u,t) \quad [2.28]$$

where $y(t)$ and $y(t-1)$ are the cost of the project until time *t* and *(t–1)*, respectively, $T(t)$ is the cost contribution of tasks that are being executed in time *t* and $R(u,t)$ is the risk management at time *t* as defined in equation [2.29]:

$$R(u,t) = \sum_{i=1}^{m} risk_i(i,t) RE_i(u,t) \quad [2.29]$$

where $risk_i(i,t)$ is 1 when the risk occurs at time *t* and 0 otherwise and $RE_i(u,t)$ is the exposure of risk R_i showing the effect of the risk *i* in time *t*, which can be calculated as:

$$RE_i(u,t) = p_i\left(II_i - \sum_{j=1}^{k} f_j(u_j)\right) + \sum_{j=1}^{k} risk_a(i,j) g_j(u) \quad [2.30]$$

where p_i is the probability of the risk R_i, II_i is the initial impact of the risk R_i, $\sum_{j=1}^{k} f_j(u_j)$ are the reductions on the impact of R_i by taking action denoted by u_j and $risk_a(i,j)=1$ if the impact of the risk R_i is reduced by taking the action *j* with a cost *j* $g_j(u)$.

Wider use of renewable energy resources encourages the use of stochastic model approaches to optimization for energy management, as discussed in [SHA 15]. Hybrid use of different renewable energy resources multiplies the risks of continuous supply. A Pareto-based approach for determining different amounts of resources to be used is a good application for stochastic processes. Integration of energy generation and storage in micro-grid applications is another field where the loads and the schedules can be optimized by using multi-level stochastic models.

Stochastic simulations have been widely used in scenario applications. In recent years, the integration of social analysis and scientific research emphasized the need for stochastic optimization. [JRI 16] is a recent paper that studies the policy implications on energy efficiency achieved by using the most recent technologies.

A simple decision model is shown in Figure 2.16. This model is inspired by solving complex industrial problems using knowledge necessary for given problem solving and knowledge of context that may influence the proposed solution. The vision, for example the amount of clean energy to be developed, is "translated" into strategies and goals to be achieved. According to goals, the actions will be decided and performed. All actions have an impact on business, environment, society and policies.

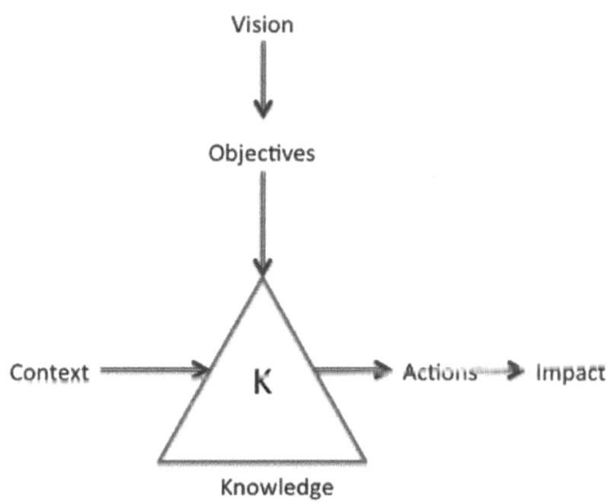

Figure 2.16. *Simple decision process [MER 07]*

The above model is a theoretical basis for designing a collaborative tool for conceptual knowledge modeling, vital in knowledge-based systems described in the next chapter.

Complexity grows with the increase in energy resources. The decision model includes both econometric and mathematical models as well as more complex techniques [NAG 00].

3

Intelligence for Energy

The methods described in the previous chapter are useful in specific situations. Energy management is a complex problem involving people, policies, businesses, the global stock of materials and the planet. Each of these stakeholders has different goals, motivations and influences. Certainly, we can use Artificial Intelligence (AI) methods and tools to help solve problems related to managing the whole lifecycle of energy; however, without considering all these factors as a whole, our solution will be partial.

The rapid increase in power demand has instigated a larger implication of technology for power production, distribution and consumption. New fields of technologies using lasers, plasma torches, superconductors, advanced materials and new electrochemistry in batteries play an important role in conservation.

After the OPEC embargo in 1973, the shortage of natural gas in 1976, coal strikes in 1978 and gasoline shortages in 1979, efficient capital has become an important issue. When energy prices increase faster than inflation, monitoring and forecasting energy prices more closely and changing the energy use patterns with detailed scheduling has become the focus of interest to avoid the high economic risks caused by energy. The conventional pricing used for electricity has hidden some investment risks [BHA 12]. Optimization methods cannot be a cure in those analyses, since the known optimization methods may not be feasible or cannot find a solution in a polynomial time [BAÑ 11]. NP-hard problems (with no solution in a polynomial time) have shown an increasing need for the use of heuristic methods, machine learning tools and AI in the energy field.

3.1. Artificial intelligence

Since 1956, AI researchers and practitioners have been inventing and experimenting with various techniques to find a solution for industrial and commercial issues. AI is above all a new way of thinking and problem-solving using all knowledge and experience that apply.

While the dream of AI founders is to build a machine that is able to act as a human, the AI interest has always been to transfer knowledge into computers and make reason using knowledge. Powered by various knowledge representations and reasoning methods, the computer in all forms is also able to learn using human-inspired and computational machine learning methods and algorithms, as well as from "observation" of the external world (robots, multi-system agents) [MER 11].

A cognitive machine such as IBM's Watson accumulates knowledge from many fields and uses natural language interface and mainly rules for reasoning. Programmed with the AI approach, a machine is able to gather all the necessary knowledge to solve a given problem and act very quickly. For example, constraint programming may be used to solve combinatorial problems with many constraints such as scheduling, planning, resource allocation and logistics; the only condition is to "know" constraints. Case-based reasoning (CBR) helps to find the right solution very quickly, only if a solution for a given problem is already registered somewhere. Exploring this analogy is very useful for matching all kinds of offer and demand.

The AI also provides techniques for speech recognition to allow communication with a computer by voice. For example, we can submit a problem to solve just by talking.

The real advantage of a properly programmed computer is that its memory is much larger than ours and its capacity of processing is quite unlimited with the latest technology. The objective, however, is to create a human–machine synergy instead of replacing humans by a machine.

The main AI techniques are as follows:

– The *expert system* contains knowledge about the field of expertise and uses the rules for reasoning. In reality, there are few situations where the rules are used to solve a given problem. This technique is effective in the areas where the rules are well defined, and there are few exceptions, such as the allocation of taxes at customs, equipment configuration, process control or in administration.

– The *cause-to-effect graph* is useful for finding a solution from an observable symptom by exploring the branches. This technique is used for diagnosis and to model the correct operation state. Building a graph for a complex system is time and energy consuming.

– The *decision tree* or static induction is used for diagnosis or in data mining. The tree is automatically generated from examples or from a prepared database [QUI 83].

– *Analogy* [BER 03, SHA 82] is very useful to learn and takes inspiration from known solutions, to build a collective experience and for any matching of supply and demand, such as demand and offer in energy, job-, travel- or house-hunting. *This technique can be used to build an idea generator.*

– *Dynamic induction* [AUR 05, MAN 94] has the advantages of induction and analogy. The decision tree is built while the user is answering the questions asked by the system. The processing of unknown values is taken into account by the analogy engine that proposes the best choice among the possibilities.

– *Constraint programming* [DIN 88] integrates a constraint propagation engine, algorithms of optimization and generation of solutions which respect the given constraints. This enables us to solve dynamic combinatory problems, such as scheduling, planning, resource allocation and management, optimization and routes or frequency allocation

– *Multi-agent systems* (MAS) are composed of specialized agents that are able to perform given tasks in interaction with other agents and the environment. Equipped with machine learning algorithms, agents are able to learn. MAS are mainly used in robots and games as well as for industrial problem solving.

The main disciplines of AI are shown in Figure 3.1. They include knowledge modeling and processing, natural language and speech processing, various reasoning techniques including non-standard logic, all machine learning methods and algorithms, hybrid systems combining knowledge representation and reasoning such as multi-systems agents, related technologies such as AI machines, multi-modal interfaces and all AI useful for Web services, including semantic search engines.

All the aforementioned techniques are useful, but their efficiency depends on the art and experience of the user in applying them for complex problem solving.

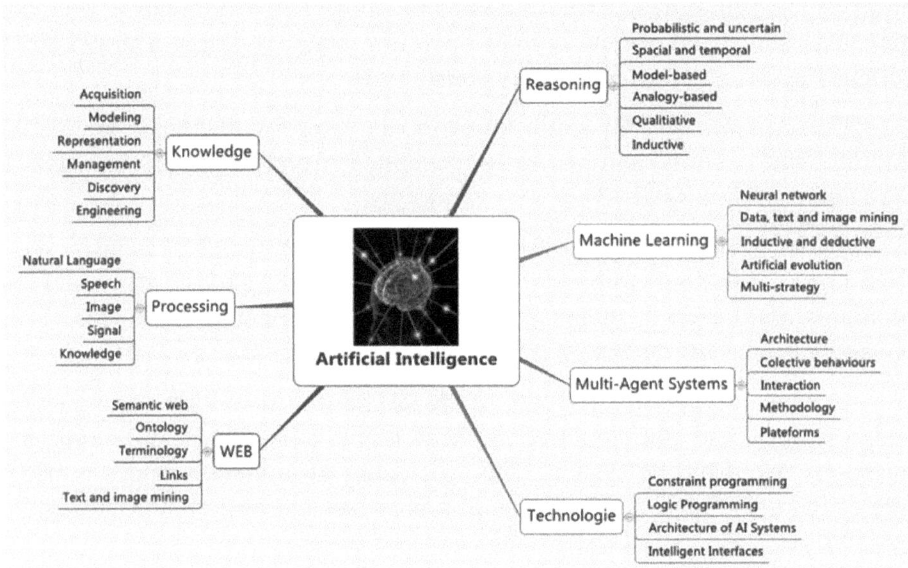

Figure 3.1. *The spectrum of AI fields [MER 16]*

In fact, solving complex problems requires the understanding of the problem in its context, the discovery of real needs and the analysis of the nature of knowledge to be exploited in order to apply the right modeling technique and decision-making mechanisms. The choice of a technique will be made as a function of all these elements. Most often, several techniques are necessary (hybrid solution).

Since the early 1980s, all AI approaches and techniques have been experimented for various industrial applications and other problem solving [RAU 85, BON 86, MER 07].

Many applications have been developed in the energy field for oil extraction such as drilling advisor, diagnosis of nuclear power plant accidents, simulations, training with e-tutor, configuration of control-command equipment, monitoring and control, equipment diagnostics, failure troubleshooting, maintenance, fuel management, reliability studies, non-destructive testing, water chemistry monitoring, network configuration and so on. They demonstrated that AI is helpful for intelligent problem solving and assistance.

Unfortunately, experiences in the energy field across the globe are not easily accessed, since people in industry do not write scientific papers. Scientific papers in the energy field are mainly based on dissertations.

3.1.1. *Expert systems*

Expert systems are among the first tools proposed by AI. The first expert system was probably Mycin, developed in 1960 for medical diagnosis [SHO 75]. The architecture of the expert system contains a knowledge model using object representation, rules for reasoning and inference engine for chaining the rules in the case of complex reasoning and dialogue with the user.

During the development phase, the expert knowledge is "translated" into hierarchically organized objects and relations between objects. In function of complexity of knowledge, meta-models can be used.

Reasoning is represented by rules. The basic form of a rule is:

if p, *then* c, *else* c1, where p represents the condition, c the conclusion and c1 another choice.

Example of a rule: *If* the energy demand is <10 MW, *then* production should increase *else* servers cannot be powered.

An inference engine selects the rules to trigger during the reasoning phase; rules use objects to reason. Objects and rules represent a knowledge base for a given domain.

The architecture of an expert system is shown in Figure 3.2. The first part represents the action of knowledge base building: a cognitive engineer discusses with one or several experts and "translates" the acquired knowledge into objects and rules. This phase is difficult, time consuming and requires validation by an expert. The design of an expert system requires a strong collaboration between expert(s) and cognitive engineers as well as choosing the right tool. Conceptual knowledge models, built collaboratively using a graphical interface, can be reused for other applications.

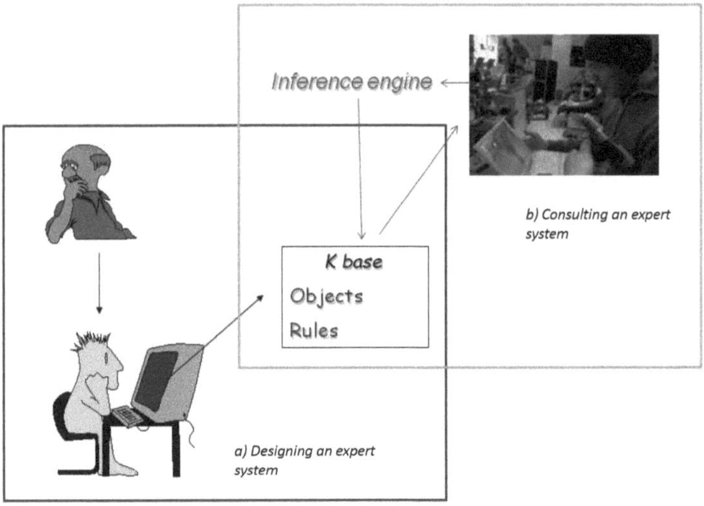

Figure 3.2. *Expert system for designing and consulting*

During the consultation phase, the user asks questions to the expert system using the interface, and the inference engine will use the rules and ask questions to the user in order to perform complete reasoning and provide the correct answers to the user. We say that the engine is "first order" when variables are used in rules. Powerful industrial generators of expert systems provide the first-order engine, meta-class facilities and reflexivity. Such generators are designed to allow full integration with other applications through information system and knowledge management.

Expert systems are suited for domains where expert reasoning is required using rules, for example in insurance, customer service, resource allocation, fault detection, failure diagnosis and so on.

In the situations where we need to capitalize the specific knowledge of several experts, it is preferable to use KADS-based acquisition tools, conceived for the blast furnace project SACHEM, where the knowledge model was built collaboratively and incrementally [DOL 96].

Energy saving has been one of the fields where expert systems are used for intensification of energy consumption while minimizing the cost of energy used by monitoring [KAL 00].

Madan and Bollinger discuss the advantages and disadvantages of using expert systems, which enable the remote monitoring, coordinating and controlling of the power distribution [MAD 97]. This early review of AI usage in the field of power systems foresees that the use of AI will spread among the researchers specifically for considering the environmental effects of power generation, transmission and use [ZWI 90].

Process industries face the problem of material, energy and other utility integrations. EXYS II is one of the initial expert system applications on the Web comparing feasible energy-saving techniques for the process industry [KAL 00]. This project helps both intensification and energy cost reduction when investing in new technologies.

3.1.2. *Case-based reasoning*

The principle of CBR is shown in Figure 3.3.

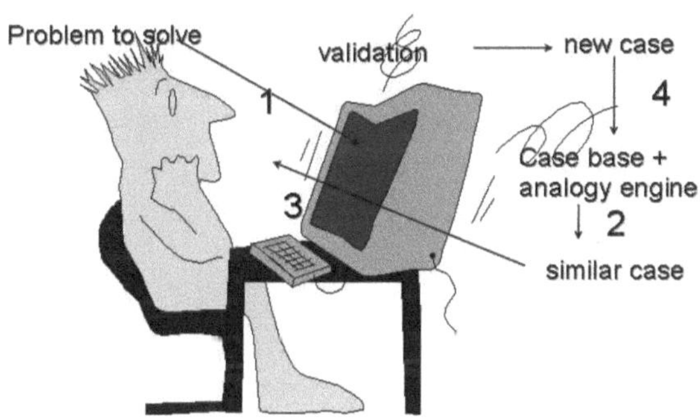

Figure 3.3. *Principles of CBR*

Knowledge in CBR is represented by cases. It can contain previous experiences, FAQ, a catalogue of products to sell or other offers. The user asks questions and then the analogy engine is triggered to select the cases to match or similar cases. It is up to the user to accept a proposed solution. If proposed cases are similar, the user decides whether one of them apply (validation phase). When a case is accepted by the user, sometimes after adaptation, it is registered in the case base and automatically available for a new consultation. We can say that the computer learned a new case.

This technique is useful to solve the problems by **analogy**. The reasoning is performed by the analogy engine.

	Problem	Param 1	Param 2	Param 3	Param n	Solution
Case 1	Light is not working	General interrupter state: on	Cable ok	Bulb hs	Connector ok	Change the bulb
Case 2	Water heater not working	General interrupter on	Cables ok	Resistance hs	Connector ok	Change the water heater
Case 3	…	…	…	…	…	…
Case n						

Table 3.1. *Case knowledge model*

Each case can be described by attributes that can be numbers, symbols, vectors and natural language description.

Three basic algorithms used by the analogy engine are induction (automated generation of a decision tree from cases – machine learning ID3 [QUI 83]), nearest neighbor (direct analogy [SHA 82]) and dynamic induction [AUR 05]. The advantage of the last one is its robustness on incomplete data.

CBR is useful for capitalization of experiences, like collective experiences in wind turbine diagnosis, for matching all kinds of offers and demand and for intelligent electronic commerce.

Efficiency is measured by the existence of analogy between cases. The cases do not necessarily need to be numerous.

3.1.3. *Constraint programming*

Constraint programming was first developed as an extension of Prolog, invented by Alain Colmeraeur's team at the Université de la Mediterranée, Marseille, France. The first constraint programming language called CHIP (Constraint Handling In Prolog) was elaborated in the European Computer Industry Research Center founded by Bull, ICL and Siemens in 1981 in Munich. Specifications were open sourced. In 1985, Bull transformed these research results to an industrial tool called CHARME. Cosytec was founded in 1990 to commercialize CHIP [COS 16]. Ilog solver was developed in 1993 and bought by IBM in 2009 as another Ilog tool – Rules.

Constraint programming was conceived for solving combinatorial problems with constraints, such as scheduling, planning, resource allocation, logistics transportation, frequency allocations for radio, television and cell phones, air traffic management , catering and crew allocation [RIV 06].

It integrates a constraint propagation engine and algorithms of optimization and generation of solutions, which respect the given constraints [DIN 88].

Constraint programming has the following fundamental principles:

– problem is fully described by listing the decision variables, parameters and constraints that limit the values of decision variables;

– solutions to such problems can be found efficiently through incremental and co-operative processes that use constraints to detect impossible combinations of values;

– constraints can be expressed by integers, Booleans, symbolic, rational numbers and real numbers.

For example, a symbolic constraint can be expressed as:

“X is a <day of the week>”

where “<day of the week> is one of <monday, tuesday, wednesday, thursday, friday, saturday, sunday>”

Typical constraints on a set of variables are:

– set S must contain the element E;

– set S1 must be included in set S2;

– sets S1 and S2 must be disjoint.

As stated in point 2, the information provided by the constraints helps to detect inconsistency fast enough to avoid an unnecessary search. This mechanism is called the *active constraint*. A constraint is not only seen as part of the problem definition, it also works as a procedural component that ensures that the relation it represents remains satisfied.

Incremental solver deducts the information about variables in the constraints. If something changes, it is able to deduce additional information from this change. This algorithm is called the *consistency algorithm*. Different consistency algorithms can be used for different types of constraints on the same problem.

Two main approaches are domain technology (two different constraints communicate through the domains of common variables) and incremental linear solver (attached to a group of linear constraints).

This constraint propagation engine reduces the space of a solution set and proposes solutions that respect constraints. Solutions can be sorted or just one best solution is proposed.

The industrial constraint programming language offers various consistency techniques; the objective is to detect inconsistencies faster to reduce the amount of additional computation and calculate the cost [CRA 92].

IBM is using these principles encapsulated in Solver to deal with various industrial problems of their clients. It is included in the "Smarter Planet" offer and applied to energy and water optimization problems.

The Association for Constraint Programming [A4C 16] organizes scientific conferences and events to discuss the progress in this field, as well as various industrial applications of this mature technique.

Arigliano *et al.* give an example of applying constraint programming to design an energy system for a manufacturing plant [ARI 14]. This open source software is accessible at [OSE 13]. With this software, connecting to the grid and energy generation using wind power, photovoltaic panels or hydrogen power is evaluated all together with storage tanks. Manufacturing plant energy requirements have to be forecasted for almost 20–25 years. The demand load is taken in a specific time horizon to fulfill the fixed load that must be met and the deferrable load responding to the less strict requirements. The system can buy and sell energy products. Designing the system means sizing all the system components and defining the energy flows through the system in a considered time period. Three scenarios are created by using the market prices of different resources and constraints defined by the plant: do nothing off the grid, penalize activities off the grid or do not penalize activities off the grid. The system gives alternatives for the technology choice (which resources should be used), sizing (what amount of each resource will be used) and dispatch (schedule the resource switches or extensions).

3.2. Machine learning

Machine learning is a group of intelligent techniques defined in 1970 by R.S. Michalski [MIC 81]; since then, many methods inspired by human learning and using computer capacity have been developed and experimented. Among them, the basic ones are AQ and ID3.

The AQ algorithm can "understand" how humans classify things, for example, if we present to someone a set of pictures on the screen representing smiling or not robots equipped with a knife or not and ask this person to "invite" by clicking robots he/she likes to his/her Club, AQ is able to recognize the rules that this person apply for selecting the robots to his/her Club (Emerald system http://www.mli.gmu.edu/papers/91-95/mli93-26.pdf).

In this case AQ "learns" the rules from the user. These rules are generalized by the learning system to serve in the other contexts. AQ can also provide explanations to the user.

The ID3 algorithm is for automatic generation of a decision tree from a large amount of examples [MIC 81]. Methods are both deductive and inductive. Later, other algorithms were added to the group of methods made for the Multi-strategy Machine Learning (MSML) [MIC 94]. The interests for knowledge discovery from data, text and image began in the early 1990s.

In contrast, the Artificial Neural Networks (ANNs) for automated learning were experimented with in the early 1940s (ANN 1.0) and used in the famous Perceptron of 1960 [MAR 09]. Later, in the 1990s, the development was stopped with ANN 2.0. Recently, with the improvement of ANN and the explosion of data produced via the Web, the boom of analytics and Big Data problems, machine learning algorithms have become very popular and are frequently used for any kind of problem. The principles of ANN are given later in this chapter. ANN cannot provide explanations to the user on "how and why" a given solution was obtained.

The range of applications of Media Server Markup Language (MSM) is very large with rule learning from examples, observations and past experiences. Data mining with induction or MSML allows real-time learning from the user or client exploring knowledge about what he/she learned, what was missing, which product he/she looked for and did not find and so on.

In the energy field, MSML can help understand how users (companies, organizations and individuals) are using energy, what amount can be spared (with propagation of effect), diagnosis, in tutor systems (with explanations), etc.

[BAR 11] uses MSML to quantify the long-term wind resource based on short-term measurements of wind resources such as wind speeds, directional distribution and turbulence intensity. Site selection or supply quantification for wind energy can be forecasted with more reality using the quantified data of wind resources.

3.2.1. *From genetic to evolutionary algorithms*

Alan Turing proposed a "learning machine", which would be parallel to the principles of evolution known in 1950 [TUR 50]. After the 1960s, many researchers worked on the idea of applying Darwinian optimization theory to computers.

Genetic algorithms (GAs) were proposed as a tool and experimented for various applications approximately in 1980 [GOL 89].

The basic principle is to generate a random population of solutions and select the best ones for further processing. These algorithms support optimization since they are used to converge for the best objective value.

Holland, Bäck and Banzhaf are the initiators of evolutionary algorithms that are defined based on genetic evolution [HOL 75, BÄC 97, BAN 98]. They are almost the same as GA but with deterministic parameters.

In GA, data is represented in genetic coding to construct the population. The aim is to improve the quality of the population using a fitness function as an objective. The algorithm is designed in four steps: [ALT 10]:

– creation of the initial population by defining the chromosomes, reproduction using simple heuristics like a roulette wheel;

– crossover operation to select the best features using simple heuristics like one point and two points to create children from the parents;

– mutation of a single gene with a certain percentage to make a change in order to avoid the blockage of local optimum;

– finishing criteria to stop the algorithm either with the number of iterations or threshold values.

There are good examples of forecasting the energy load [HOO 05] demand for different kinds of energy resource [AZA 15] using GA. Since 2005, GA has been the preferred optimization method implemented in technology or policy design [FON 10]. Furthermore, this method is easy enough to be implemented in energy-efficiency and performance calculations [REN 10].

GA used in forecasting is given in the sample application of Kaucic [KAU 09]. In this study, the focus is the excess return of the energy industry by the EU index for different countries in Europe. The researcher takes 16 variables in finance, globalization, inflation and technical groups, which are represented in a chromosome to find the econometric models to be applied for the 12 months ahead under prediction constraints. The algorithm tries to converge for the best combination of countries in terms of economic knowledge input through the chromosomes.

3.2.2. *Multi-agent systems*

Multi-agent systems (MAS) were born from the convergence of two disciplines: automation and AI.

The main principle is shown in Figure 3.4. Agents are specialized for performing a given task. They may be cognitive or reactive. The execution of given tasks is supervised by a controlling agent.

Current MAS include a variety of agents implementing the main AI techniques and are used in robotics, games and for knowledge discovery, among others.

The application of MAS in the energy field can be used for the energy management of distributed power sources, in energy grids, etc. More information is available on the website of the IEEE Power & Energy Society Multi-agent Systems Working Group [IEE 16].

MAS can be programmed to find a better solution for a given problem.

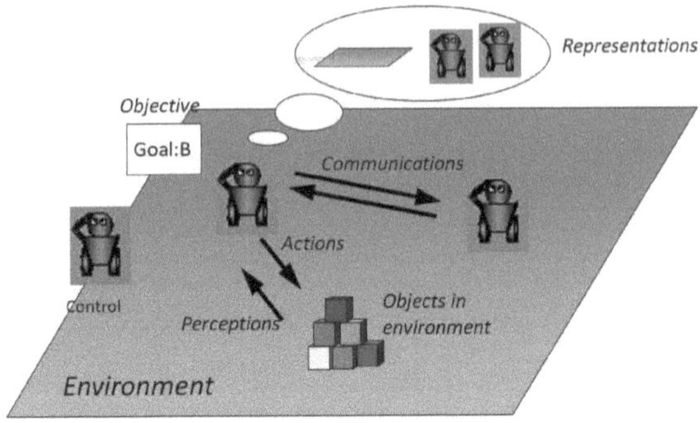

Figure 3.4. *Basic principle of MAS [FER 95]*

3.2.3. *Collective intelligence*

Collective intelligence proposes algorithms inspired by the observation of a population of bees and ants for simulating the intelligence usage by a group of individuals. Ant colony and particle swarm optimization (PSO) are the most commonly known algorithms [PEN 06].

In the ant colony algorithm, a population of ants is initiated with the possible or random solutions where each ant is assigned a transition probability. Then, the pheromone trail is updated. At each iteration, the pheromone amount is evaporated and the best ants can survive using the following equations defined by Goss [DOR 99].

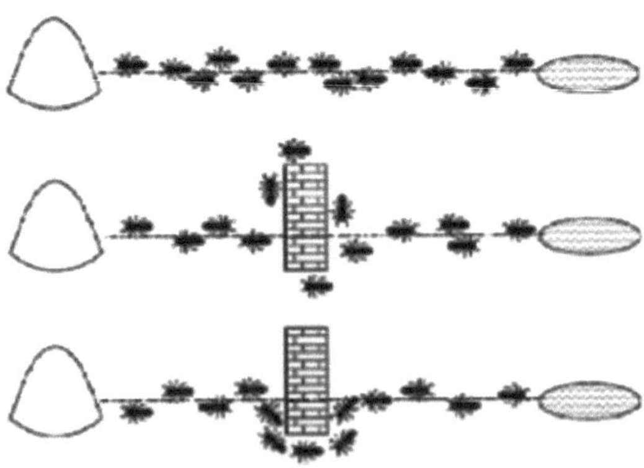

Figure 3.5. *Typical food search of ants [GHA 13]*

Probability with heuristic information η and S, and constants α and β depends on the ratio of the individual pheromone τ_{ij} at iteration t to the total of τ_{ij} for all the ants:

$$p_{ij}(t)=\frac{\tau_{ij}(t)^{\alpha}\eta_{ij}^{\beta}}{\sum_{j\in S}\tau_{ij}(t)^{\alpha}\eta_{ij}^{\beta}} \quad [3.1]$$

where pheromone is renewed by the evaporation rate ρ:

$$\tau_{ij}(t+1)=(1-\rho)\tau_{ij}(t) \quad [3.2]$$

The trail is renewed for the global best path $\Delta\tau_{ij}$:

$$\tau_{ij}(t+1)=\tau_{ij}(t)+\Delta\tau_{ij} \quad [3.3]$$

A number of energy forecasting studies are performed using the ant colony algorithm [DOR 06, SHA 15, TOK 09].

PSO has the unique property of finding the individual best (particle best – Pbest) as well as the social best (global best – Gbest), simulating the flocking bird groups [SHA 15].

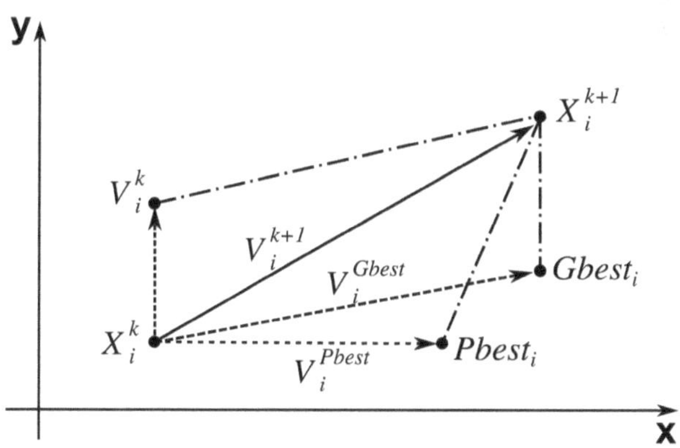

Figure 3.6. *PSO geometry in finding the particle best and the global best [KEN 95]*

Given the position and the velocity as below, the objective function is calculated for each particle to find the position and velocity for the particle best and the global best using the random number r:

$$x_{ij} = x_{\min} \times r(x_{\max} - x_{\min}) \quad \forall i = 1,\ldots,N; \forall j = 1,\ldots,n$$
$$v_{ij} = \alpha \frac{x_{\min} + r(x_{\max} - x_{\min})}{\Delta t} \tag{3.4}$$

After having found the particle best and the global best, the velocity and positions are recalculated using the constants c_1 and c_2:

$$v_{ij} \leftarrow v_{ij} + c_1 q \left(\frac{x_{ij}^{pb} - x_{ij}}{\Delta t} \right) + c_2 r \left(\frac{x_{j.}^{sb} - x_{ij}}{\Delta t} \right) \quad \forall i = 1,\ldots,N; \forall j = 1,\ldots,n$$
$$x_{ij} \leftarrow x_{ij} + v_{ij} \times \Delta t \tag{3.5}$$

Kitamura shows a good example of a two-objective operation planning problem for a period of 24 h in industry using PSO, with the objectives being the simultaneous minimization of energy costs and CO_2 emissions [KIT 06]. An operation schedule is worked out by dividing the problem into 1 h intervals.

Clustering-based application of PSO helps scheduling to be performed in fewer of iterations. It has applications in battery charging or reservoir loading optimization. Events are grouped according to the efficiency loss or filling time and the best is chosen in one example.

PSO is widely used in forecasting and clustering in the energy field. Kumar *et al.* [KUM 13] give a good example of optimizing the power flow in the grid. In their analysis, the total load of electricity is optimized by considering the busses, the branches, the generators and the transformers.

3.2.4. *Artificial neural networks*

The ANN is a technique that mimics the human brain. The ANN is limited to connections between artificial neuron and automated learning from data presented to the input of the ANN. In 1943, Warren McCulloch and Walter Pitts created a computational model for neural networks based on mathematics and algorithms called the threshold logic. This model opened the way for research on neural networks; however, multi-layer definition of the ANN started the second fame of this algorithm by the end of the 20th Century. Neural networks were used in the first known AI machine.

"A neural network is a massively parallel processor made up of simple processing units, which has a natural propensity for storing experiential knowledge and making it available for use. It resembles the brain in two respects:

1) data is acquired from the environment through a learning process run in the network;

2) interneuron connection strengths, known as synaptic weights, are used to store the acquired knowledge" [SAL 98].

The ANN can be defined as a directed simple graph that will contain at least one start node, one end node and at least two nodes representing the process elements with a state variable, a weight and a bias associated with each node. At least two processor elements are connected in parallel. A flow on each link (beginning and process or process and end notes) is exactly the same as the output caused by the link. A supervised, non-supervised or reinforced learning algorithm helps to model the desired output [GUR 11].

It needs a differentiable activation function, which is, in general, one of the sigmoid, tangent, hyperbolic or Gaussian functions. The most widely used sigmoid function is expressed as follows:

$$f(x) = \frac{1}{1+e^{-x}} \quad [3.6]$$

One of the initial examples of using ANNs in fault detection is the 1990 implementation of Hoskins and Himmelblau on chemical reactors [HOS 90]. A network with three input nodes, four hidden nodes and five output nodes is trained with observations of values of fault reasons. The changes are noted and the output given for each reason is followed if it shows any faulty state. This model is adopted for a nuclear basis later on. An ANN model designed for quality control of the coal used in electricity generation is one of the most interesting models with 9 input, 4 hidden and 18 output nodes [SAL 98].

The feedforward ANN is preferred for forecasting, signal flows, evaluation of binomial classes and time series analysis. In this type of learning, for each node n_j, neural input is taken as the weighted sum of all the flow towards it and a bias:

$$n_i = \sum_k w_{ki} x_k + \theta_k \qquad [3.7]$$

Therefore, in the sigmoid (logistic) function, the x variable is replaced by the summation of flows. The root mean square error (RMSE) is the most widely used error besides the mean square error. It is expressed as the total of squared difference of the supervising value and the calculated value:

$$E = \frac{1}{2} \sum_k \left(y_k^s - o_k \right)^2 \qquad [3.8]$$

where y_k^s is the supervising value for the kth output when o_k is calculated.

Back propagation (BP) is the most widely used supervised training method for multilayer feedforward networks [CIN 10b]. It is a gradient descent algorithm, which improves the resulting performance of the network by reducing the total network error. This is done through adjusting the weights using the differential chain rule. During BP learning, weights are modified according to their contribution to the error function by using the learning rate η, giving the rate of change at each iteration:

$$\Delta w_{ik} = -\eta \frac{\partial E}{\partial w_{ik}} \qquad [3.9]$$

When the forecasted value is compared with other methods, the relative error can be used, which normalizes the mean square error by dividing the achieved value with the sum of the squares of the supervising values.

If there is a big variety of random initial values using a different combination of input parameters, BP might have the disadvantage of local convergence.

A typical application of BP could be in prediction, classification or clustering.

A good example is given for power voltage prediction in multiple output when the load demand is a time series used for the input [NAG 10].

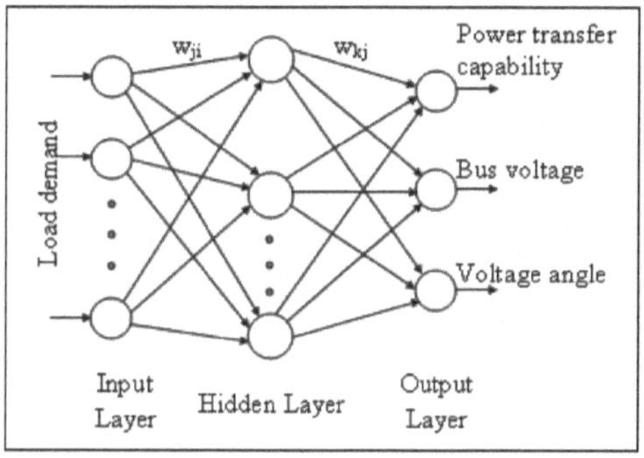

Figure 3.7. *A typical BP application for predicting the power voltage [NAG 10]*

Self-Organized Map (SOM) is an unsupervised learning method designed by Kohonen for fault detection [KOH 90]. The goal of self-organizing maps is to convert a complex high-dimensional input data into a simpler map. Thus, SOMs are appropriate for cluster analysis, where underlying hidden patterns among records and fields are sought. The SOM structures the output nodes into clusters of nodes, where nodes in closer proximity are more similar to each other than to other nodes that are further apart. This technique is based on competitive learning, where the output nodes compete among themselves to be the winning node, the only node to be activated by a particular input observation. The output nodes would have scoring values using a function, the most commonly Euclidean distance between the inputs and weights.

The ANN using unsupervised learning is called SOMs, designed by Kohonen [KOH 90]. SOM is a topologically ordered network (mostly two-dimensional) used to extract knowledge about the nature of the input data. It works on the competitive learning that choses the winning neuron according to similarities.

Each neuron is defined with a vector of features **m** and a position **g**. The SOM works in two steps. In the first step, the competition is run among the input neurons, which is given by:

$$WinningNeuron: c = \arg\min_i \left\| x - m_i(t) \right\|, i = 1,2....,n \qquad [3.10]$$

In the second step, the group of winners is taken in a neighborhood:

$$m_i(t+1)=m_i(t)+\alpha f_{ci}(t)[x(t)-m_i(t)] \quad [3.11]$$

where α is the learning rate (which can be taken as the standard or as changing at each iteration) and f_{ci}(t) is the activation function, which is usually Gaussian defined between the winning neuron and any other neuron.

The specialty of this approach is that both the closest unit and its neighbors move towards the input.

In Figure 3.8, an example of node clustering is given with only three inputs, namely the X coordinate, Y coordinate and energy level, that could be used for any network where the energy level of the nodes is important. There are no historical data available, and clusters are differentiated not only with the differences in data distances, but also with the centroid of clusters being apart.

This method is widely used in clustering or mapping. [RÄS 08] is a good example of clustering the electricity used by the households to find the possibility of energy saving.

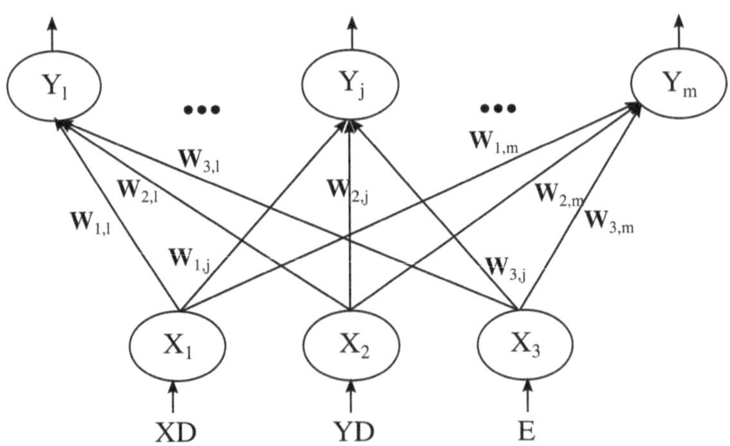

Figure 3.8. *Energy node clustering with three inputs [ENA 10]*

Domingues *et al.* [DOM 11] collected data on the electricity consumption of buildings from power systems. The SOM is trained by combining the meteorological data with the collected consumption data to train the SOM. The analysis reveals the active and reactive electricity consumption of buildings during different tariffs of the grid. Comparing the cluster reactions with peak hours, the study suggests a new price tariff in that city.

Reinforcement learning is mainly used in robotics defining the states; therefore, a majority of the applications use Markovian decisions since it has to be in states and that is why it is widely used in robotics. A typical application in the energy field can be shown by the Memarzadeh example applied in wind farms, focused mainly on the uncertainty of the wind strength that defines a 5-tuple Markovian decision chain (S, A, T, R, γ), with the following assumptions [MEM 15]:

– S is a finite set of states of the system;

– A is a finite set of possible actions that can be taken, which affect the evolution and rewards;

– there is a transition probability function that models the uncertainty in the prediction of the next state, given the current state s_t and action a_t; the conditional probability for the next state is the transition:

$$T(s,a,s^t) = P(s_{t+1} = s^t \mid s_t = s, a_t = a) \qquad [3.12]$$

where s is the vector of states and a is the vector of activities.

The Markov property holds to satisfy that the current state depends only on the previous state so that future states are independent of the past.

$R{:}S \times A \rightarrow R$ is a reward function based on the current state s_t and the action a_t. It can represent an immediate cost (negative reward).

$\gamma \in [0, 1)$ is a discount factor that discounts the future rewards and relates them to the present value.

These processes are iterated from state to state with a policy, π: $S \rightarrow A$.

The policy can be a probabilistic function like Bellman or Boltzmann [ZHU 12]. For the uncertain environments with renewable energy resources, it is a preferred type of learning.

3.2.5. *Bayesian networks*

The Bayesian approach is at the crossroads between mathematical probability and the concept of inference borrowed from AI.

Bayesian networks are also called naive Bayes, belief networks or probabilistic networks with graphical methods.

Bayes theory, invented by Thomas Bayes in the 18th Century, has an important effect on statistical inferences today with the big data concern. This theorem gives the probability of a cause as its effect on the result. That is why it is expanded to use in the cause-and-effect analysis. It may depend on the experiences of experts (belief network) or on the historical probabilities observed [CIN 10a]. The use of Bayesian networks has been widely accepted due to their ability to handle the missing data. This method can use background knowledge and experimental data to complete each other. If the probabilities of events in association are known exactly, the results are closer to the true results [DAI 13].

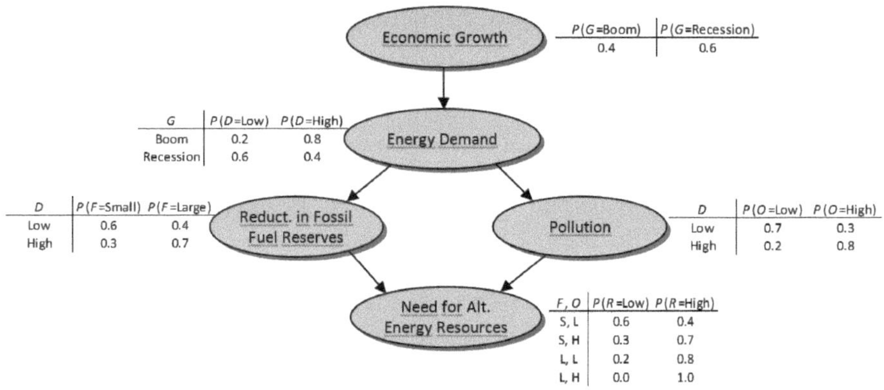

Figure 3.9. *A simple example of a Bayesian network [CIN 10a]*

Bayesian networks form a directed acyclic graph, where nodes represent random variables of interest and edges represent informational or causal dependencies among the variables. The conditional probability table constitutes the quantitative part of the network with conditional probabilities of nodes. These probabilities rely on the chain rule, which is about the joint probability distribution of each variable.

According to the chain rule, the marginal and conditional probabilities can be computed for each node of the network. For a set of variables $X_1,...,X_d$, the joint probability of X_i is then given by:

$$P(X_1,...,X_d) = \prod_{i=1}^{d} P(X_i|\text{parents}(X_i)) \quad [3.13]$$

When evidence is received from external sources about the possible states of a variable or a set of variables, the marginal and conditional probabilities of the variables can be computed by marginalizing over the joint. If some evidence is given over some variables, the probability of incident of some events is calculated as follows:

$$P(U|e) = \frac{P(U,e)}{P(e)} \quad [3.14]$$

where U is the universe of variables $X_1,...,X_d$.

Ordering of the variables affects the results in the chain calculation. It is recommended to use a topological ordering by which the variables are ordered such that every variable comes before all its descendants in the graph [KES 12].

3.2.6. *Support vector machines*

Vapnik pioneered the development of support vector machines for classification into two classes. This method aims to develop an optimal hyperplane as a decision function using the maximum distance between the two class vectors. The support vector uses nonlinear mapping for the input vectors. A high-performance decision plane is developed to divide the training data into two classes. The distance among the closest data on each side of the hyperplane is expected to be large enough to have a better performance in classification. The support vector machine can be expressed as an optimization problem in primal and dual representations as shown by Ercan and Kayakutlu [ERC 13].

The primal optimization problem is defined as follows:

$$\begin{aligned} &\min z = \frac{1}{2}\|w\|^2 \\ &s.t. \\ &r_i\left(w_i x_i + w_0\right) \geq 1 \end{aligned} \quad [3.15]$$

where w is the weight vector, w_0 is the bias and r_i is the class label that can take a value of −1 or 1. The simplest output function dependent on the weights is expressed as:

$$y = \sum_i w_i x_i + w_0 \quad [3.16]$$

The support vector machine is designed to differentiate one output as −1 or +1, whose (w_i, w_0) couple has to satisfy the below constraints:

$$w_i x_i + w_0 \geq 1 \; while \; r = 1 \quad [3.17]$$

$$w_i x_i + w_0 \leq -1 \; while \; r = -1 \quad [3.18]$$

(x_i, r_i) data are called the ***support vectors*** if they satisfy the equality form of the constraints:

$$r_i\left(w_i x_i - w_0\right) = 0 \quad [3.19]$$

(x_i, r_i) data are called the ***support vectors*** when the equality form of the constraint is satisfied. This is the case when the distance between the support vectors and the optimum hyperplane is calculated using the equality equation. The norm of the weight vector is expressed in terms of $r_i \subset \{-1,1\}$, and the difference from zero on the positive or negative part is expressed as ρ:

$$\frac{r_i(w_i x_i + w_0)}{\|w\|} \geq \rho_i \quad \forall i \quad [3.20]$$

The smallest distance on both sides of the optimal separating plane will give the distance between the support vectors and the optimal hyperplane, and by considering $\rho\frac{1}{|w|}$, the total margin will be at least $\frac{2}{|w|}$.

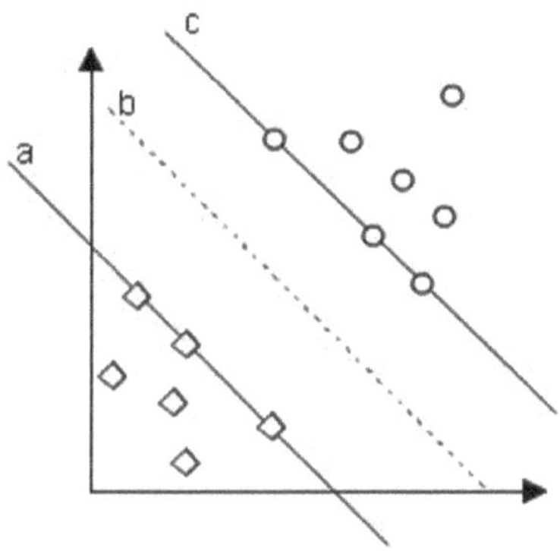

Figure 3.10. *Support vectors separating the peak power prices that fall and those that rise [PAP 14]*

The primal problem has a nonlinear objective function and a single linear equality constraint that can be relaxed into a linear problem by using Lagrange multipliers:

$$L_p = \frac{1}{2}\|w\|^2 - \sum_i \lambda_i r_i (w_i x_i + w_0) + \sum_i \lambda_i \qquad [3.21]$$

where λ_i are Lagrange multipliers with $\lambda_i \geq 0$:

$$f(w) = \frac{1}{2}\|w\|^2 = \frac{1}{2}\sum_i w_i^2 \qquad [3.22]$$

The convex function $f(w)$ is continuous and twice differentiable [BER 99], where $\nabla^2 f(w)$ can be easily computed to give a positive-definite identity matrix.

By converting the problem to its dual and using the regularization parameter C as a penalty factor, the Kernel function $K(x_i,x_j)$ can be derived for $0 \le \lambda_i \le C, i=1,2,...,N$. The best parameters are chosen according to the kernel function:

$$\max z = -\frac{1}{2}\sum_{i=1}^{N}\sum_{i=1}^{N}\lambda_i\lambda_j r_i r_j K(x_i,x_j) + \sum_{i=1}^{N}\lambda_i$$
$$s.t.$$
$$\sum_{i=1}^{N}\lambda_i r_i = 0$$
$$0 \le \lambda_i \le C, i = 1,2,...,N \qquad [3.23]$$

The Kernel function is also named as radial basis function by machine learning experts.

The error achievement desired and the standard deviations showing the width of the radial basis function should be defined before application.

This method is widely used for fault detection of energy generation equipment, performance measures of operations, renewable energy performance predictions or market forecasts, as proposed in [PAP 14].

3.2.7. *Non-standard logics*

Solving complex problems in the industrial world requires a toolbox for describing a typical knowledge, situation or state that cannot be expressed by traditional methods. In forecasting, the collected data change quickly and lose value over time. In knowledge-based systems supporting this activity, it is vital to express the value of data in time. A temporal or non-monotonic logic may be used. In the case of a blast furnace, the experts state that "when the color of flame is between orange and violet the operator should activate parameter 321". Here we can use an uncertain or fuzzy logic. Logic of beliefs may serve to express a point of view or intuition. Other logics are conceived to deal with very specific situations. A variety of complex systems can be found in the energy field.

As one of the most recent techniques, we detail the fuzzy logic and limit the description of others as a specific logic.

3.2.8. *Fuzzy logic*

For optimization problems with uncertainty, either stochastic optimization or fuzzy logic can be used for solution. The main difference between crisp and fuzzy sets is that an element belongs to a certain set with a degree of membership μ, and the membership of a function in the given set is defined by $\mu(x)$, which is a value between 0 and 1 [ZAD 65]. The nearer the value of membership to 1, the higher the belongingness for the range. When the input data is defined in a fuzzy set, triangular or trapeze representations can be used. A sample of fuzzy logic will be given by explaining the fuzzy goal programming, where fuzzy goals are characterized by defining the tolerance limits and the aspiration levels of constraints [DAI 10]. For a weighted goal programming the problem of G, the number of goals with weight w_g membership is to be maximized [BAR 99]. To explain the weighted goal programming model g, the number of goals is minimized or maximized with weight w_g. So that the λ variable is defined as:

$$\lambda = \Sigma_g w_g \cdot \mu_g(x) \text{ where } \Sigma_g w_g = 1 \qquad [3.24]$$

λ is maximized in the final model, aiming at maximization of the belongingness proportional to weights.

The final model is expressed as:

$$\begin{aligned} & \textit{Maximize} \quad \lambda \\ & s.t. \\ & \qquad \lambda \le \mu_g(\vec{x}) \\ & \qquad C_h(\vec{x}) \le b_h \\ & \qquad \vec{x} \ge 0 \end{aligned} \qquad [3.25]$$

Let L_g represent the lower limit which is equal to the aspiration level, U_g represent the upper limit which is equal to the sum of the aspiration level and the upper tolerance limit on the right-hand side of the goal, and $y_g(x)$ any data calculated. For a minimization goal, the membership is defined as:

$$\mu_g(\vec{x}) = \begin{cases} 1 & y_g(\vec{x}) \le L_g \\ \dfrac{U_g - y_g(\vec{x})}{U_g - L_g} & L_g < y_g(\vec{x}) < U_g \\ 0 & y_g(\vec{x}) \ge U_g \end{cases} \qquad [3.26]$$

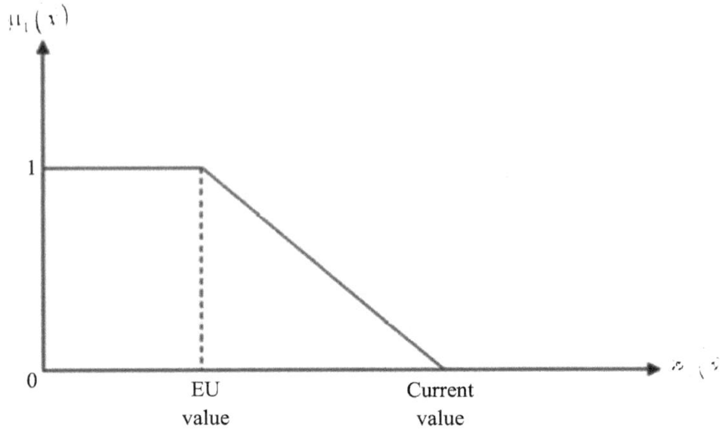

Figure 3.11. *Minimizing the total CO_2 based on the EU limit [KAY 15]*

Figure 3.11 shows an example of minimizing CO_2 emissions. L_g represents the lower limit which is now equal to the difference between the aspiration level and the lower tolerance limit, U_g represents the upper limit and $z_g(x)$ expresses the interval on the right-hand side. If the problem has a goal of maximization, the membership will be defined as data-lower limit.

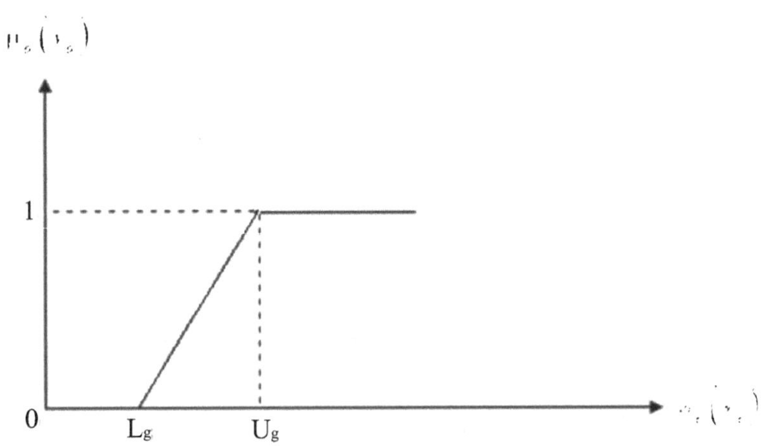

Figure 3.12. *Linear membership function for a maximization goal*

3.3. Hybrid algorithms

Strategic decision processes for energy portfolio investment and/or renewal and extension of the existing systems need further accuracy in the prediction of lifecycle costs and efficiency improvements. The complexity of global competition causes more realistic long-term demand projections and short-term resource, demand and price forecasting. Hybridization of techniques is preferred to monotonic methods when dealing with conditions of vague knowledge, no knowledge or randomization for assumptions. Hence, the intelligent methods are improved to be hybrid, dynamic and adaptive.

3.3.1. *Hybrid architecture for industrial problem solving*

Real-world complex problems cannot be solved with only one technique. While the academic approach is to apply the newly invented technique to every kind of problem, the right approach is to understand the problem first, understand the context of the problem and how far the proposed solution has to consider the links and propagations.

An industrial application has to communicate with all other related applications. An industrial application has to be conceived as a part of a company information system and connected with all concerned components of this system.

For example, a decision support system for talent management needs at least four communicating components:

1) a database (or case base) of available resources described by capabilities, expertise and knowledge level;

2) a case-based system for matching demand, offers and training;

3) a scheduling system (CP) for long-term resources planning, short-term allocation and training;

4) real-time scheduling (expert system able to find the required and available resource corresponding to the needs).

Combinatorial problems sometimes require hybridization between constraint programming and other optimization techniques such as simulated annealing, operational research or heuristic algorithms. [RAG 15] is a good example of redesigning the production schedules based on the energy demand at peak hours. Each schedule is defined with a GA using the early and latest start times of the jobs on the machines considering the time window, where the jobs that occur in the peak energy usage time construct a set of constraints.

Each intelligent technique has its own specific capabilities. Some are better at learning, others at avoiding the blockage of local optimum and some at exploring the decisions. That is why the complex world of energy needs a combination of intelligent techniques in order to create synergy for solution. People focused on optimization first looked for the efficient combination of diversification (exploration) and intensification (exploitation). Therefore, simulated annealing (SA), tabu search (TS), GA, and evolutionary programming (EP) have been widely used combined in a group or inter-search algorithms.

Iwamatsu shows one of the earliest applications in atom clustering to minimize the energy landscape by combining the conjugate gradient method with the evolutionary algorithm [IWA 02].

Iwamatsu defines an N atom system with an optimum arrangement for the global minimum of the total potential energy *U(x)*, defined by:

$$U(x)=1/2\sum_{i=1}^{N}\sum_{j=1}^{N}v(r_{ij}) \qquad [3.27]$$

where v(q) is the pair potential (Lennard-Jones potential) that represents the interatomic interaction between two atoms i and j in the cluster.

$$v(r_{ij})=4\varepsilon\left(\frac{1}{r_{ij}^{12}}-\frac{1}{r_{ij}^{6}}\right)$$

where [3.28]

$$r_{ij}=\sqrt{(x_i-x_j)^2+(y_i-y_j)^2+(z_i-z_j)^2}$$

is the distance between two atoms i at (x_i, y_i, z_i) and j at (x_j, y_j, z_j). ε is the material parameter that can be fixed as 1.

The cluster is minimized using the conjugate gradient method and extreme values that increase the total value of energy use are eliminated. The most current atom clustering sample uses the GA with the particle swarm algorithm [WAN 15].

Multi-objective consideration of cost minimization in parallel with carbon emission minimization has also been one of the reasons for the hybridization of different methods. Nonlinear and non-continuous functions can easily be handled by the multi-objective evolutionary algorithms that integrate the GA and PSO. An example is seen for a thermal system design for cogeneration plants. Fazlollahi *et al.* [FAZ 14] investigated the effects of energy system sizing and operation on CO_2 emissions. They tried to minimize the total cost of operations and carbon taxes and minimize the carbon emissions using the decision variables of maximum available capacity for the technology used. Binary variables are used for technology choice to define the master problem of mixed integer programming and the slave program of emissions. Finally, the multi-objective evolutionary algorithm and the Pareto-optimal solution are applied.

The most widely used technique is the GA with the fuzzy inference (FI) or ANN. This is mainly because the random population used by the GA helps to optimize the parameters required by other systems before the implementation starts. As in the case of the ANN, it is a big advantage to optimize the learning rate, the number of nodes in the hidden layer, the momentum rate and the initial weights before running the model.

A good example of using ANN algorithms for learning and then as a fit function for the GA can be found in heat storage [KAN 10]. Another field where ANN is preferred is energy load forecasting. Hong used a support vector and after the classification of data, he estimated the parameters of the ANN by using the GA [YU 15].

Site selection for offshore wind turbines currently a problem based on the maximized wind power and the minimized noise effect [LEE 10]. The optimization problem was defined to maximize the energy density (P) that satisfied the criteria of the maximum h and maximum distance (R) from the coastline. P is characterized by using the following function:

$$P=\frac{1}{2}\rho\int_{0}^{\infty}V^{3}f(V)dv \qquad [3.29]$$

where ρ is the density, V is the instantaneous wind speed and $f(V)$ is the Weibull distribution. Nonlinear productivity of the ANN is used for the spatial calibration of longitude and latitude showing depths.

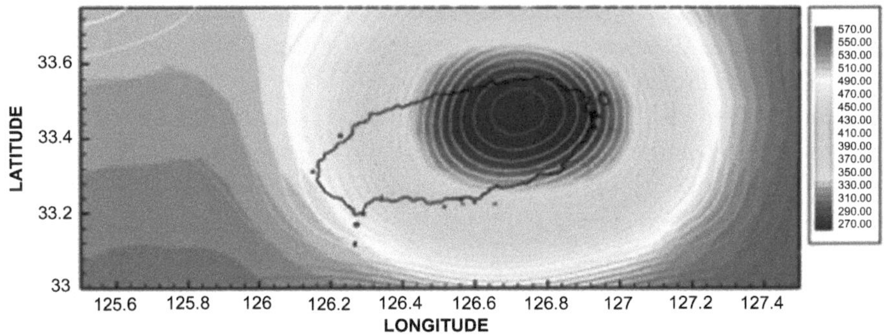

Figure 3.13. *Candidate locations drawn by ANN depth diagrams [LEE 10]. For a color version of the figure, see www.iste.co.uk/kayakutlu/energy.zip*

Interestingly, the hybridization of the ANN and GA is used for short-term load forecasting, where researchers take the error function produced by the ANN as the fitness function for genetic chromosomes [WAN 10].

The fitness function is given in a logistic function as:

$$f(x)=\frac{1}{1+E(x)} \quad [3.30]$$

$$E(x_s)=\frac{1}{2n}\sum_{i}^{n}\sum_{j}^{c}(y_{ij}-o_{ij}^{s})^2 \quad [3.31]$$

where the error is calculated for a population of size n and crossover of c on M candidates for the sample s. The threshold is the difference between the expected value for the ith input and the jth output and the output value calculated for the sample s.

It is also very common to optimize the ANN parameters using PSO as in the very short-term wind forecasting [PRA 11].

3.3.2. *Adaptive intelligence methods*

Adaptive intelligence has been developed for robotic improvements. In complex problems, there are a lot of uncertainties and nonlinearities. Sometimes, uncertainties are not solvable using stochastic approaches, but fuzzy mathematics is needed. If the problem is not convex and not differentiable, then it is almost impossible to solve the nonlinear function. One of the initial examples was the kinematics problem, given in the couple of x and y positions with joint angles θ_1 and θ_2 [HAS 06]:

$$\begin{aligned} x &= l_1 \cos\theta_1 + l_2 \cos(\theta_1 + \theta_2) \\ y &= l_2 \sin\theta_1 + l_2 \sin(\theta_1 + \theta_2) \end{aligned} \quad [3.32]$$

It is difficult to define those angles in inverse sine and cosine expressions with the distance to the angles given in l:

$$\begin{aligned} \theta_2 &= \cos^{-1}\frac{x^2 + y^2 - l_1^2 - l_2^2}{2l_1l_2} \\ \theta_1 &= \tan^{-1}\frac{y}{x} - \tan^{-1}\frac{l_2 \sin\theta_2}{l_1 + l_2\cos\theta_2} \end{aligned} \quad [3.33]$$

Correct angles and distances need to be found for the solution. Hence, adaptive learning is used in ANN applications where the calculated angles and distances are used as input to the unsupervised system.

The hybrid use of fuzzy systems with ANN, called the adaptive network-based fuzzy inference system (ANFIS), is a widely used intelligent solution for unsupervised networks with no historical data. The ANFIS uses the feedforward BP learning that has five layers:

Layer 1: fuzzy membership functions, where the fuzzy expressions are processed with each node being adaptive;

Layer 2: fixed nodes representing an output of a rule fired using a fuzzy minimum operator;

Layer 3: reserved for the normalization process to scale the firing strengths where the nodes are fixed;

Layer 4: adaptive nodes operate as a function block of linear combination using the weights in second and third layers.

Layer 5: sum of outputs of layer 4.

BP is applied on the results of layer 2 and the least square error is calculated on the results of layer 4. The only user-specified information is the number of member functions.

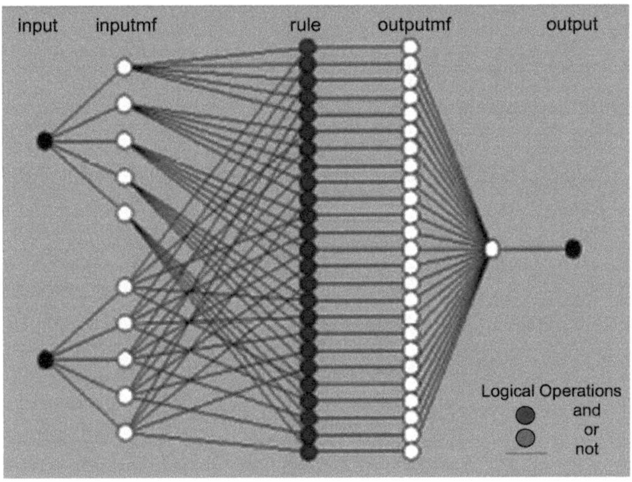

Figure 3.14. *An example of an ANFIS structure*

A good example of ANFIS use is given in pipeline performance measures, where a thermodynamic system is measured using an intelligent system [MOH 14]. In this sample, adaptive nodes are represented by membership functions for the adaptive rule set A and the input x with adaptation premises x^* and σ as follows:

$$\mu_{A_i}(x) = e^{-\left(\frac{x-x^*}{\sigma^2}\right)} \quad [3.33]$$

In the second layer, the outputs are just the multiplication of membership values and the input is:

$$O_{2i} = \omega_i = \mu_{A_i}(x_1) \times .... \times \mu_{C_i}(x_n) \quad [3.34]$$

The third layer is the normalization where the output becomes:

$$O_{3i} = \varpi_i = \frac{\omega_i}{\omega_i + \ldots + \omega_n} \quad [3.35]$$

The output of each node is calculated as:

$$O_{4i} = \sum_i^n \varpi_i f_i \quad [3.36]$$

where f can be taken as a fitness function, which is Gaussian in the example:

$$O_{4i} = \sum_i^n \varpi_i f_i \quad [3.37]$$

The last layer is the sum of all the outputs achieved in the fourth layer:

$$O_{5i} = \frac{\sum_{i=1}^{i=n} \varpi_i f_i}{\sum_{i=1}^{i=n} \varpi_i} \quad [3.38]$$

Gas transmission network optimization for the pipelines is defined by using the GA in order to find the best fitness value to be used in the ANN [MOH 14].

Bayesian adaptive ANN is used when both prior and posterior probabilities are required. Bayesian law of depending on prior events can be used to take weights from a probabilistic distribution and calculate the posterior probabilities [ALP 11] as follows:

$$P(A|B) = \frac{P(B|A)P(A)}{P(B)} \quad [3.39]$$

Bayes theorem describes the way in which the assumption about observing the event A is updated by having observed the event B. In Bayesian inference, the prior and the likelihood are mathematically combined to produce the posterior. The posterior is then used to determine the probabilities of specified effects. For Bayesian estimation, this theorem is often expressed as [LI 12]:

$$p(\theta|x)=\frac{p(\theta)p(x|\theta)}{\int p(\theta)p(x|\theta)d\theta} \qquad [3.40]$$

where $p(\theta)$ is the probability density of the prior distribution for the parameter(s) of interest, $p(x|\theta)$ is the probability density distribution of the posterior for data x given θ, and $p(\theta|x)$ is the posterior density of the distribution of θ given data x. The normalization of coefficients ensures that the integral of the posterior is always equal to 1. In the sample article, the wind speed is estimated using the Weibull distribution and ANN errors are calculated with the estimated and standard deviation values.

The following strategic and operational energy related functions can be improved by using intelligent techniques: effective control of equipment and services, real-time data collection on the resource availability and air pollution; service center management for effective micro-grids; short-, medium- and long-term forecasts of market prices and demand; macro-level models for each country.

Automated demand response (ADR) is a new dynamic approach used in intelligent energy programs. It was initially used in smart buildings, but later improved for smart grids playing several roles: users, metering, distributer, regulator, pricing and third party service. ADR programs can make interruptible rates possible or enable load shifting. ADR offers significant opportunities to save energy and optimize electric grid performance.

A good example of improvements in long-term forecasting is shown by IEA. IEA has created different scenarios by considering various hybridization, tax and price changes to analyze the trends in 2035, where the industrial energy use seems to drop down to 29.4% [INT 14a]. Figures 3.16 and 3.17 are extracts from multiple scenarios to show the forecasts of the share of fossil resources and the share of different energy users in 2035, respectively.

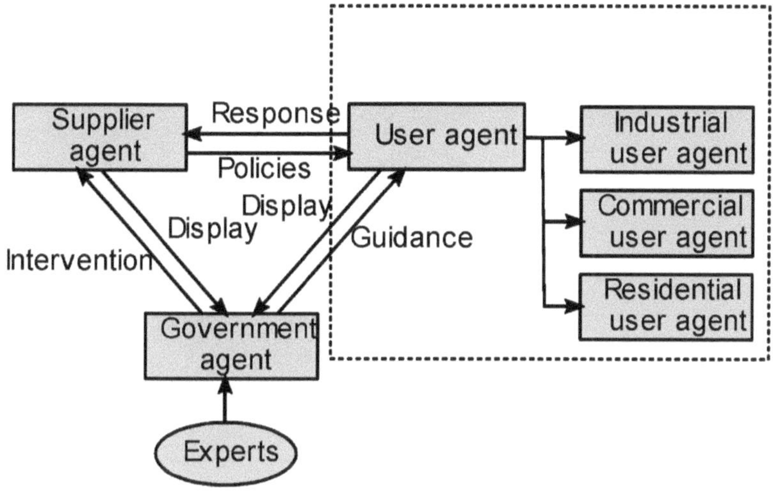

Figure 3.15. *Multi-agent scheduling example of ADR [YAN 15]*

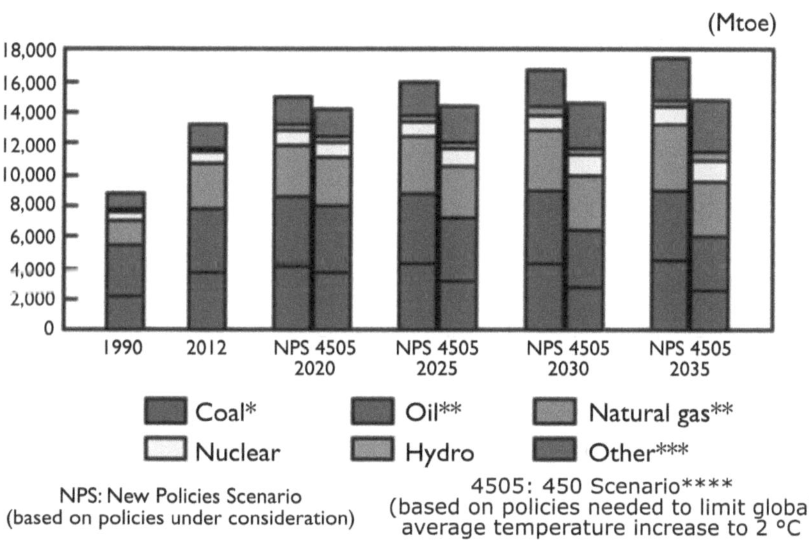

Figure 3.16. *IEA future trend analysis by energy resource until 2035 [INT 14b]*

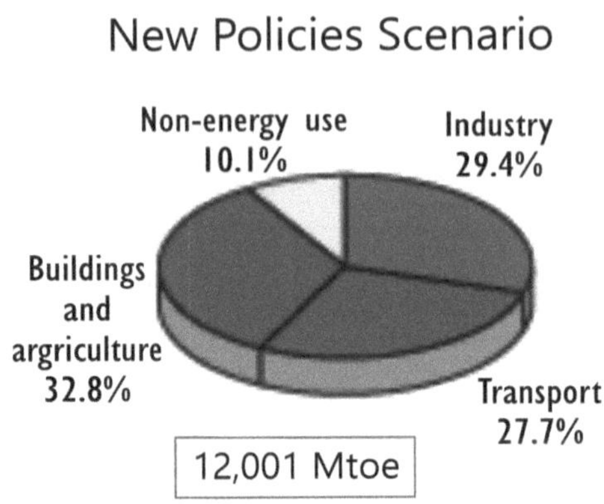

Figure 3.17. *IEA future trend analysis by sector until 2035 [INT 14b]*

The IEA forecasts the intensification of renewable energies in all sectors. This phenomenon results from the combination of at least three factors: awareness about planet condition, governmental incentives in some countries and perspective of business. Renewable energies are not new, but their extensive use can be considered as innovation and contributes to energy greening.

However, the progress should be made in their intelligent generation and frugal use for the benefit of humanity. The technologies presented above may help achieve this ambitious and necessary goal under the condition of learning the art of wisely using them: “right technologies for the right problem”.

The next chapter presents the main trends, actions and experiments aiming at more intelligent use of energy.

4

Greening and Innovating with Renewable Energies

After initial warnings in the 1960s, a second more recent wave of alerts about global warming have raised awareness of the depletion of fossil energies and have initiated international summits aiming at defining goals and fostering actions in favor of alternative energies and water protection to halt climate change and environmental pollution. The objectives of greening have always been to reduce CO_2 emissions by various actions, energy saving and social innovation aiming at supporting entrepreneurship to reduce poverty. Most of the innovations in energy that followed international decisions concern the use of renewable energies and the discovery of new sources of energy. This chapter focuses on the interactions between greening enterprises, economic models and innovation.

4.1. Agenda 21 to COP21

This interest in greening is not new. These debates have always been strongly influenced by the contradictions of trade and environment. While some countries headed their actions individually, global governance is still missing because international negotiations have to face the conflicting interests of the participants. The first global conference was organized in Stockholm in 1972 by the United Nations (UN) and brought together the representatives of 113 countries. In 1983, a World Commission on Environment and Development put emphasis on the risks for the future of mankind in the Brundtland Report [COU 87]. Restrictions to productions and consumptions to reduce the degradation of the ozone layer were stipulated in the Montreal treaty in 1989 by the agreement of 195 countries. Rio 1992 followed all the contradictions and became the starting point for sustainability efforts. Individual governmental acts were monitored in the 10th and 20th year meetings. Finally, COP21 changed the agenda for real commitments.

4.1.1. *Rio 1992*

International efforts in greening started in the early 1970s but took longer than a decade to formalize. The UN brought together 178 government representatives in 1992 to propose a global plan for a sustainable development in a meeting called "Earth Summit I" in Rio de Janeiro, Brazil. Originally, there were three main subjects to discuss: saving the forests, limiting the release of polluting gases to slow global warming and conserving biodiversity to save plants and animal species. Well before the meeting, the first subject (forests) was abandoned because of high budget needs for compensation of lack of income in the Third World countries. The opening speech presented mentioned the aim of breaking the vicious circle of poverty.

There were two UN reports opened to signature "Framework Convention on Climate Change" and the "Convention on Biological Diversity". The negotiation on avoiding desertification also began. Agenda 21, the main output of Rio Summit, was the first attempt for a plan of action to promote greening; hence, it became the starting point for sustainable growth. Governments agreed to gather and share information on greenhouse effects, national policies and best practices. They gave commitments for launching national strategies to avoid emissions, which also included economic and technological support for poor countries to prepare for the impacts of climate change.

This was a reunion triggering local initiatives for global solutions. Five years later, 130 countries met in New York to review the progress since the first summit. A 10-year plan with models and methods was suggested. The more ambitious targets were only lightly touched on by the developed countries and not taken into account at all by the poor countries.

After the Earth Summit in 1998, the UN General Assembly explicitly identified climate change as a concern for humanity. However, the Assembly did not agree on binding measures to push the governments to take action. No global institution was established, and there was no encouragement for changing the laws to protect the environment. The tools which would enable the realization of the Rio 1982 declarations were governed by the biggest economies and were not internationally available [PAL 92]. The general public was totally ignorant about the existence of such a declaration [DER 13].

As a follow-up for the UN meeting, the first model to analyze the impacts of climate change on economic development was defined in 1992 by the Dynamic Integrated model of Climate and the Economy (DICE) [NOR 95]. Nordhaus [NOR 95] analyzed in this model the geography, temperature and income together and worked on the effect of temperature changes versus income changes.

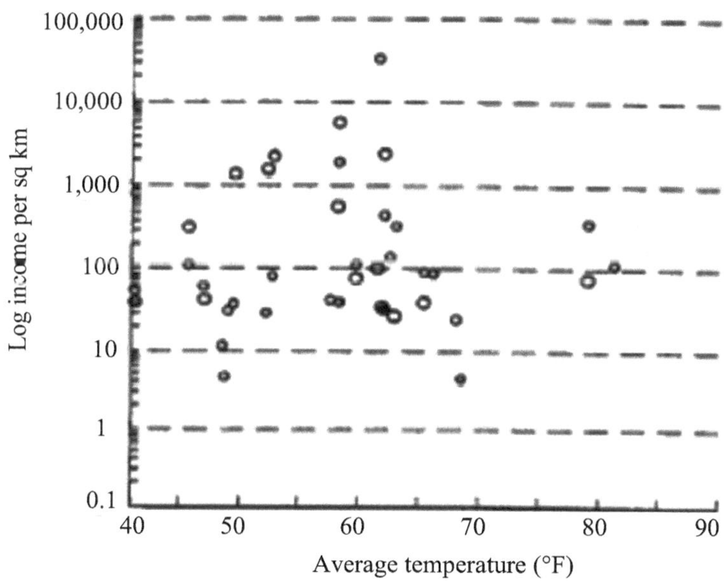

Figure 4.1. *Temperature and income in the DICE [NOR 95]*

The results obtained through studies of richness and temperature increases showed big gaps in different economies as a result of global warming. He indicated that climate models resembled macroeconomic models. This was the first scientific analysis to show the negative impacts of global warming, stating that CO_2 increases in parallel to rising temperatures. He showed the effect of 2.5° warming as 10 times the volume of carbon in the air. Yet he observed uncertainties in improving the richness of certain geographies and opened a philosophical discussion: "should human betterment should be before nature". Nordhaus invited social scientists and economists to work together on the uncertainties and concerns caused by global warming. DICE models developed on Excel by Nordhaus became a tool for governments to observe the warming effect in their region.

The damages are calculated per industry and as an impact of weighted output or weighted population, as shown in Table 4.1.

	Output weighted	Population weighted
Agriculture	0.13	0.17
Sea level rise	0.32	0.12
Other market sectors	0.05	0.23
Health	0.10	0.56
Non-market amenities	–0.29	–0.03
Human settlements and ecosystems	0.17	0.10
Catastrophes	1.02	1.05
Total	1.50	1.88

Table 4.1. *Damages of 2.5° warming on industries as a percentage of output [NEW 10]*

This impact analysis was the beginning of new economic models, but caused a lot of debate among scientists.

One of the opponents, DiBona, claimed that the theoretical potential of energy efficiency up to the maximum can be defined by the physical laws which can be redefined for both power and heat as follows [DIB 96]:

– technical potential offered by innovation;

– economic potential based on the private internal value;

– economic potential based on the social value;

– effective governmental policies to avoid failures or barriers.

4.1.2. *Kyoto 1997*

The Kyoto meeting of 1997 was one of the positive effects of Agenda 21. December 1997 was an important landmark for the developing countries: the Kyoto Protocol was signed. It focused on CO_2 emissions with a commitment of 5% reduction on 1990 levels, until 2008–2012. One of the highest polluters, the USA, did not commit but proposed to trade the emissions on the international basis. Nordhaus developed the Regional Integrated Model of Climate and the Economy (RICE). Using RICE 1998, he calculated the cost of the Kyoto Protocol until 2100 as $716 billion, two-thirds of which is to be borne by the USA [NOR 99]. This was one of the calculations that kept the USA away from sustainability commitments.

The UN elaborated a report on a 5-year overview of improvements after the Earth Summit. In that report, it was clearly stated that poverty was reduced in transition countries but that the global economy caused negative effects on a growing number of poor countries [UNI 97].

4.1.3. *World Assessment of Resource Management model*

Carraro and Galeotti designed the World Assessment of Resource Management (WARM) to show that sustainability efforts do not avoid the economic development and that tax is not the only instrument to avoid emissions [CAR 97]. WARM covered the equilibrium of the energy demand among 12 European countries (1978–1989) for a trade balance. The model included research and development to lead innovation of new energy supply technologies, end-use technologies that directly cause efficiency increase, technologies that directly alert demand of energy and new consumption technologies changing the energy use and indirectly alert energy demand for production. The model allowed labor demand wage bargaining as well as the diffusion of the technologies.

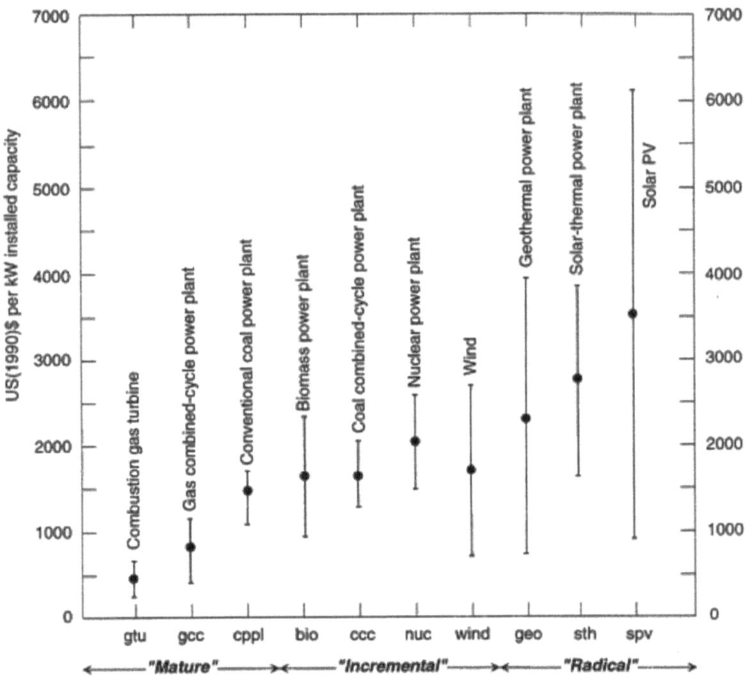

Figure 4.2. *Energy technology versus installed capacity [GRU 99]*

The impacts of energy generation, use and trading are analyzed for three major players: firms, households and governments. Econometric models are constructed by considering the energy markets, labor markets and international trading in Europe with the rest of the world. The technological impact is emphasized in this model [EUR 95]. Experiments of WARM applications showed that the short-term taxes should be reduced on wages, but it will not be important in the long term.

4.1.4. *PRIMES model*

The PRIMES model focuses on learning curves for electricity-producing technologies that do not cover the spillovers or the research and development failures [GRU 99]. The economic model developed by Capros and Kokkolaki, aims to create and use several scenarios in economic calculations [CAP 96]. It is used by a larger number of European countries.

These bottom-up models were first applied in Denmark with success [JAC 01]. The model determines an equilibrium by simulating the prices under demand uncertainty. It incorporates standard energy policy issues, security of supply, strategy, costs and environmental issues, standards on new technologies and renewable sources and policy issues regarding electricity generation, gas distribution and new energy forms. The model is constructed by different energy production systems and demands of the buying sector where some demanders may also be suppliers.

PRIMES was the first model to include both the bottom-up (engineering, explicit technology choices) and top-down (microeconomic foundation of economic decisions by agent) elements for the supply demand equilibrium in energy markets. PRIMES 2 was improved in 2001 and used for the analysis of Global Energy Outlook by International Energy Agency (IEA).

4.1.5. *Market Allocation model*

The first widely accepted bottom-up analysis of energy technologies was Market Allocation (MARKAL), in development since 1980 [SEE 02]. It was designed to minimize the total cost of a market consisting of primary resources, heat and power generation, delivering to residential and industrial users, taking into consideration the demand together with other constraints.

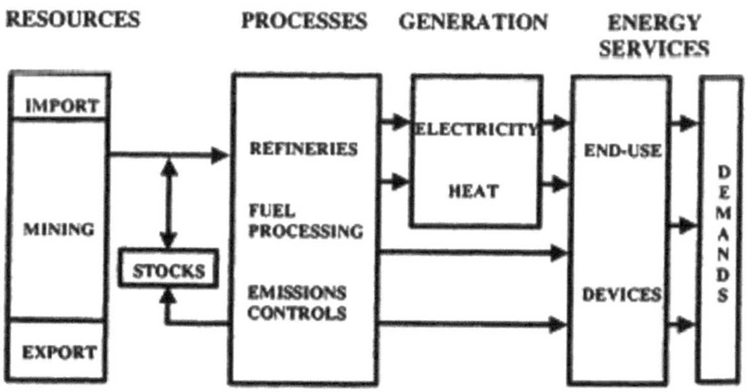

Figure 4.3. *MARKAL market components [SEE 02]*

In 2001, MARKAL was applied by the Energy Technology Systems Analysis Program (ETSAP) of the IEA using a linear programming model. Nonlinear programming parts would be included in the following years.

Energy supply and demand curves, costs and benefits of different technologies, emission profiles, energy supply chain and system costs are input for the system. The model is run on thousands of constraints to identify the least-cost energy systems and to identify effective responses to emission boundaries. This macroeconomic model tries to evaluate new technologies as well as taxes and regulations. It gives a prospective analysis of long-term energy balances and tries to evaluate the regional cooperation.

This optimization tool provides a new vision on energy price instruments (e.g. tax and carbon emission permits), energy technology policies and economic growth calculations at the macro-level. The framework for market-based calculations avoids the speculations besides allowing the allocation and reallocation of research budgets. The MARKAL model is also seen as a supporter of developments in solar and wind power investments. As an output of the model, equilibrium prices and quantities of energy, material and environmental commodities are achieved.

This model can be used by an enterprise, a country or by inter-trading countries at the same time. Economic strategies suggested by the system are to be executed linked to the environment, and the energy economists must well analyze the output.

More details can be found in [OPE 16]. Training on MARKAL is given by the IEA [ETS 16].

4.1.6. *Johannesburg 2002*

The initial Rio 1992 excitement fell like a soufflé. After the summit, the participants and governments discovered political and economic barriers. The USA, the most important contributor to the disastrous situation, had not undertaken the actions decided by the collaborative commitment. The need for a new meeting grew with reactions.

The 10th year overview of the Rio Summit was realized in Johannesburg, South Africa, in 2002. Under the heading of "Sustainable Development", a wide range of subjects were discussed to include poverty, water quality and availability, cleaner energy, health, good governance, technology, production and consumption, oceans and fisheries, and tourism. In his opening speech, Kofi Annan declared:

> "Let us stop being economically defensive, and start being politically courageous. And let us face an uncomfortable truth: the model of development we are accustomed to has been fruitful for the few, but flawed for the many. A path to prosperity that ravages the environment and leaves a majority of humankind behind in squalor will soon prove to be a dead-end road for everyone." [UN 02]

A declaration with 37 items was announced at the end of the meeting, where the main decision was that, to achieve the international goal of sustainability, more effective and accountable "international and multilateral" institutions would be needed [LEP 14]. Commitment to the UN declarations and to monitor progress at regular intervals was emphasized.

It is recognized that the social consciousness for sustainability is growing but improvements to protect the natural world are moving at a slow pace. Almost all the UN members joined the conference except the USA. This was the first meeting where it was recognized that globalization is supporting the growth of wealth gap among the rich and the poor. It was recognized that 20% of the population in the highest developed countries account for 86% of total individual consumption, whereas the poorest 20% account for only 1.3% [BRU 02]. As a result of this recognition, economists started working on a new macroeconomic model.

The Extended Energy System (ERIS) was developed by the International Institute for Applied Systems Analysis (IIASA). This bottom-up approach discussing the learning curve in energy innovation was suggested and applied globally. It defined the learning rate through a learning index used in knowledge stock functions. It focuses on different heating systems and technologies comparing developed and underdeveloped countries [BAR 04]. The project showed that the learning by doing method, preferred for solar energy technologies, enforced more global collaboration than the learning by searching method, preferred for fossil energy resources. The overall aim of research and development in the field of energy was to maximize utilities of the new technologies for each individual by contribution of the sustainability and efficiency research.

Endogenous technological change in the DICE Global Warming Model (ENTICE) was developed by Popp to show that research has diminishing returns on energy consumption but the effect of innovation on both the price and savings occurs immediately [POP 04]. This research clearly states that the following barriers should be avoided for continuing innovation:

– technological factors depending on the basic sciences;

– government policy and regulations;

– cultural and psychological factors;

– infrastructure and maintenance;

– undesirable social and environmental effects.

4.1.7. *The Integrated MARKAL-EFOM System model*

In 2003, the UK announced governmental policies expecting a 27% reduction of energy use by reducing heat and power demand by improving building quality: replacing equipment with energy conserving appliances and installing local, clean and energy-efficient power supplies [KEL 06].

Efforts for The Integrated MARKAL-EFOM System (TIMES) project started in 2000, which brought improvements on the bottom-up market-based approach by combining the technology approach with the economic approach. It was developed in the IEA ETSAP group to eventually replace the MARKAL model. The project considers the technologies, the commodities and commodity flows according to different scenarios. A new efficiency calculation for transportation, residential and commercial/ industrial demand sectors was introduced [BLE 07].

The policy scenarios used by the TIMES were constructed by considering the economic, technological and environmental impacts linked to the gross domestic product (GDP) and capital requirements. Some examples of those scenarios include nuclear energy to be phased out completely, a considerable decline in coal usage, subsidizing and reimbursement for the investment in renewable energies and efficiency improvement programs for residential buildings. Efficiency improvement is expected to occur through thermal integrity and more rational use of energy through awareness programs.

Strong incentives for cogeneration and more efficient technology use were recommended for energy generation. This model tries to find the best equilibrium for the supply and demand, once all the input is provided and the scenarios are defined. The demand is defined as its driver's elasticity power [LOU 08].

The main improvement made by the TIMES is not to take the fixed demand. Changing demands under constraints such as 40% renewable energy penetration and 20% emission reduction are met with minimized cost of supply. Linearity of the model makes it simpler to use. Power and heating prices, emission amounts, technology capacities and even abatement costs are among the outputs.

Moreover, energy commodity prices, energy flows, marginal emissions and even abatement costs can be calculated. It is enhanced to have a multi-country and global analysis.

More details can be found in [OPE 16]. Training on MARKAL is given by IEA [MAR 16].

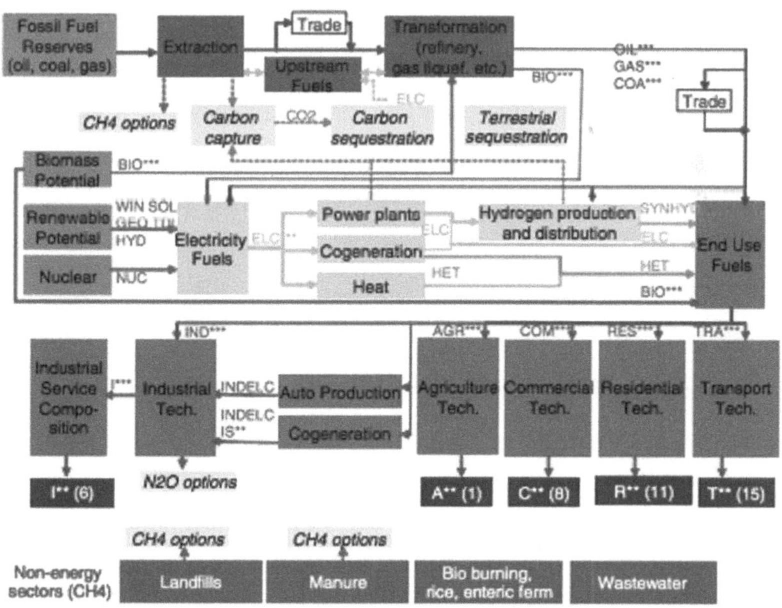

Figure 4.4. *Renewable energy approach using the TIMES model [LOU 08]. For a color version of the figure, see www.iste.co.uk/kayakutlu/energy.zip*

Between 1998 and 2005, a 25% efficiency improvement in Germany was foreseen by the application of the TIMES model.

The EU Renewables Directive (RES) came into action for EU members in 2009. The motivation of the directive was to reduce greenhouse gas emissions, promote the security of energy supply, promote the technological development and innovation as well as provide opportunities for employment and regional development in rural and isolated areas [LIN 13]. The TIMES model was applied to Norway to implement the EU RES directives. It demonstrated that the green certificate market would contribute wind investments in Norway. The Kyoto protocol was renewed in Stockholm in 2009, again with a limited number of countries contributing. European countries showed high interest following the EU directives.

In 2015, 79 institutions in 38 countries were using the MARKAL and TIMES models.

4.1.8. *An example of country use: UK MARKAL model*

The University College London Energy Institute developed the UK MARKAL model in 2010 [USH 10]. In 2008, the Committee on Climate Change reported that the UK must significantly reduce greenhouse gas emissions by 34% by 2020 and 80% by 2050. Adaptation of the MARKAL model to this report was realized with the following changes on the model:

– a more detailed treatment of CHP was included and some data for the service sector and residential were changed;

– new constraints were added to impose feasible technologies;

– bioenergy chain was redesigned to include biomass;

– additional flexibility and building technologies were defined;

– model recalibration was made to include non-energy emissions and to handle medium-term uncertainties;

– near-term policy announcements were made on renewable energy new coal-fired capacity.

A good example of a combined application considers both the global approach with its international considerations and industry-specific local conditions. The TIMES allows scenario definitions and the MARKAL allows specific data input. As such, we can come up with a decision of keeping the old and new technology together where the improvement of building energy performances, microchips and solar energy is promoted [COM 12].

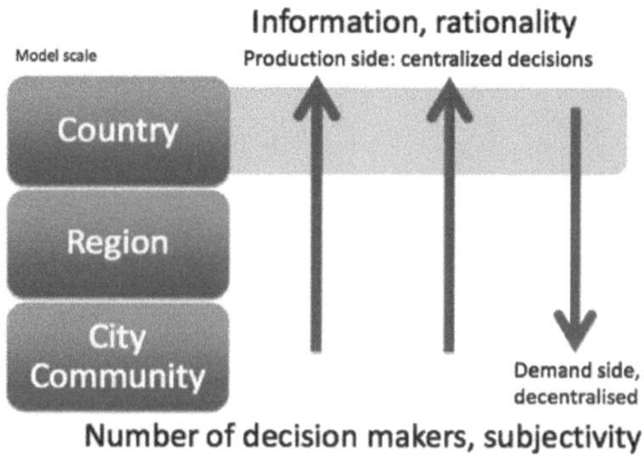

Figure 4.5. *Framework of the SOCIO-MARKAL model [KAN 10]*

4.1.9. *Socio-MARKAL model*

A social dimension is added to the MARKAL model with the support of IEA-ETSAP. This approach considers the technological contributions combined with the behavioral impacts. They suggest that it is important to collect data on the willingness/resistance of the inhabitants, and their environmental awareness or ability to identify the drivers of greening in the society, as shown in Figure 4.5 [NGU 11].

This model flows in the following steps:

– Hypothesis generation: qualitative research to identify potentials of behavioral change regarding energy consumption. The qualitative expressions are based on data that includes semi-structured interviews, observations and social experiments. The energy-saving benefits are considered without a reduction in performance. The essential character of behavioral change must be clearly explained. If the interviewees are interested in campaigns, the awareness program is designed to reduce the lack of information and motivation, the cost of changing technology, etc.

– Hypothesis testing: survey research to test and measure hypotheses generated based on questionnaires, and rank and sample statistical analyses realized in the first step.

– Behavioral change scenario process: construction of long-term scenarios including behavioral change, in particular expert-built scenarios from the collected data.

– Design: transformation of the SOCIO-MARKAL data and scenarios to feed the MARKAL database.

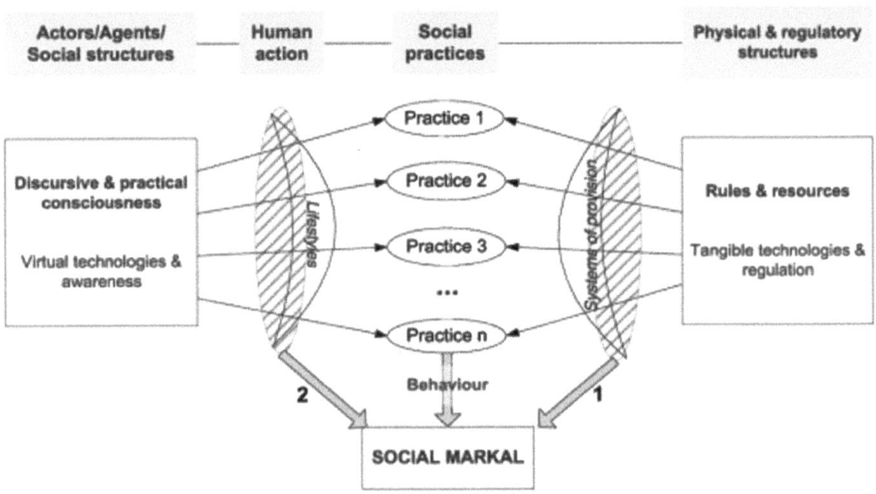

Figure 4.6. *SOCIO-MARKAL concepts [NGU 11]*

This model suggests implementation of awareness analysis of change in energy technologies and awareness in behavioral change as the knowledge increases. In order to launch the SOCIO-MARKAL model, barriers and benefits to depict the awareness should be defined and strategies will be designed to be implied in the hypothesis [NGU 11].

4.1.10. *Improved PRIMES model*

A new version of PRIMES was defined by the National Technical University of Athens to include the subsector-specific energy equilibrium. Subsector-based demand-side definition and increasing the energy resources to 17 main sources have been the major improvements.

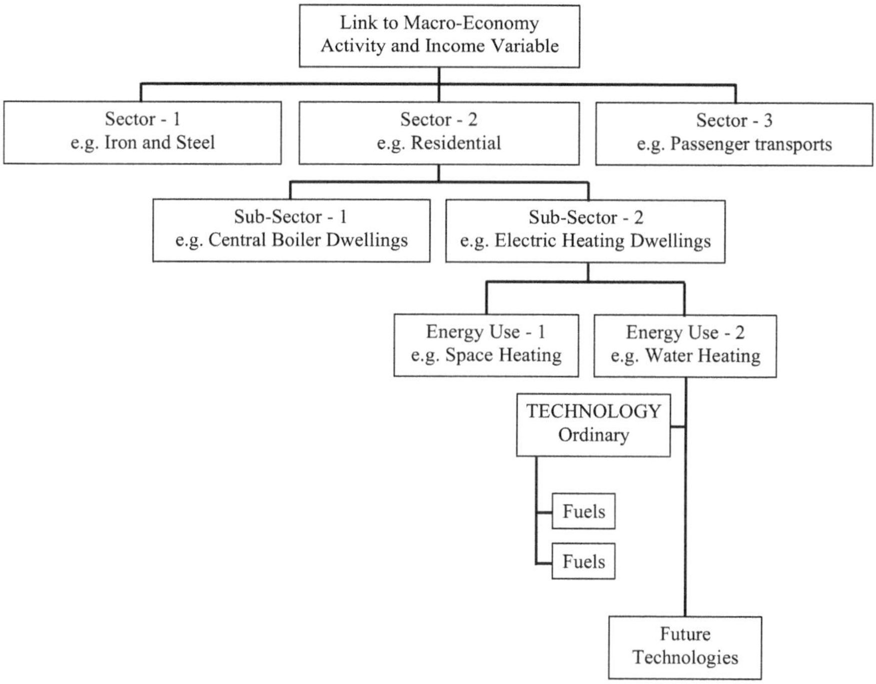

Figure 4.7. *Demand-side subsectoring in PRIMES V2 [CAP 12]*

4.1.11. *Rio 2012*

Economic market models implemented in lots of different countries should be analyzed and compared at the international level; that is why the amount of greening is put on the table in Rio, Brazil. The Rio meeting was repeated 20 years later on June 2012, when it was observed that only four of the 90 initial objectives of Rio 1992 showed significant progress. There were 283 commitments registered in Rio 2012, including topics such as poverty, sustainability, economic development, trading and monitoring of the commitments [RIO 12]. The aim for the future is stated as an "economically, socially and environmentally sustainable planet" where humans are at the center of sustainability. The countries represented in 2012 reaffirmed to strengthen the international collaborations and political changes to avoid poverty and to secure the environment. A framework for institutionalizing sustainability work and action-based plans was constructed. There were good and bad traces over the 20 years. CO_2 emissions had increased by 38% with the major part occurring between 2002 and 2010 [SCH 12]. It was stated that forest reductions had declined and ozone-damaging substances reduced. The most pleasant news was of course the exponential increase in renewable energy investments.

Strategic models for energy optimization have been developed specific to application systems such as buildings, transportation and cogeneration/tri-generation systems [CAN 14].

The Sustainable Development Summit of September 2015 in New York defined the action plan of the following 15 years to have a better world in 2030 [GEN 15]. The assembly ended with 17 sustainable development goals and 169 targets concerning people, planet, prosperity, peace and partnership.

Decision support models are being developed for integrated management of all the utilities including power, heating, cooling and water. Transportation and buildings are included in the management system, yet the tool is to be used by operators not necessarily equipped with the necessary energy knowledge [GIR 15].

Distributed energy generation and use has become the main subject with focus on smart grids. The year 2015 has taken a place in energy history as the year of radical changes in energy markets.

4.1.12. *Towards ecological macroeconomics*

During the same week of Rio 2012, an international conference on ecological macroeconomics was held in Rio. Conference discussions focused on Peter Victor's book *Managing without Growth* [VIC 08]. The book presents five main simulation models using system dynamics:

– a model for Canada when low or zero development occurs;

– a model when there is only a limited contribution of less energy consumption to well-being;

– a simulation model of energy and economics for the USA, considering that the peak production of oil has already occurred;

– a model for examining the scope of changes required in high-, medium- and low-income countries required to respond to a specific CO_2 emission boundary.

The details of the models are presented in [PVI 08].

This research field made the socio-economists and technologists work together. Steve Keen's "Debunking Economics" model [KEE 10] and Campiglio and Bernardo's "Modelling the Transition" [CAM 12] were the initiators of ecological macroeconomic models. These models are drawing attention to the financing of green products and the conservation of nature by proposing new credit and debt models.

4.1.13. *COP21 Paris-France, 2015*

Just before the Climate 2015 meeting, oil prices became a real issue for oil suppliers with the USA reducing imports and Iran entering the Western market as a new supplier. The demand of OECD countries has decreased, while the demand of non-OECD countries like China and India has grown. Clean energy supporters became suspicious that investors would go for cheap fossil sources. Oil-related industries had revenue declines.

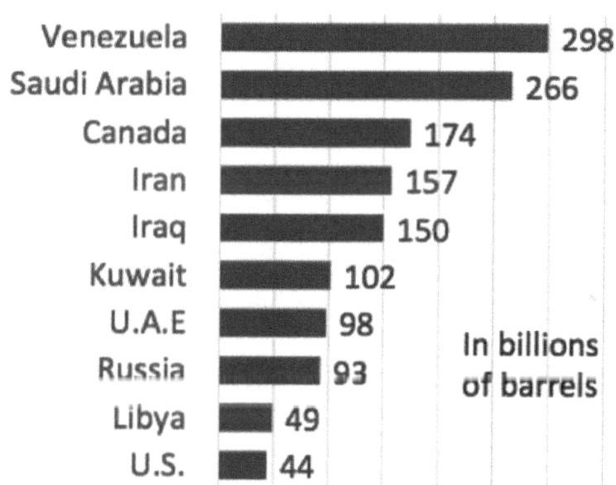

Figure 4.8. *Proven oil reserves [DAL 15]*

However, energy culture is growing. Industries have started installing energy conservation equipment and started moving plants for less transportation. Mass transport systems are promoted by smart city projects and the power supply chain is enriched to give better services to consumers. China slowed down development and India gave more pace to highway constructions. Renewables were used to generate 90% of electricity in Europe. Carbon emissions were flat.

In December 2015, 195 countries signed the first universally legal climate agreement. In that agreement, a global action plan was defined to keep the temperature increase below 2°C and rapid reductions in emissions using the best available scientific tools. The attendees were committed to be transparent, to support developing countries and strengthen the ability of societies to deal with the impact of changes. Fatih Birol, the current President of the IEA, stated that COP21 is a historical milestone to stimulate innovation in the energy sector [HAR 16]. He also stated that IEA is changing to involve non-OECD countries like China and Mexico to realize the target of becoming a global clean energy hub.

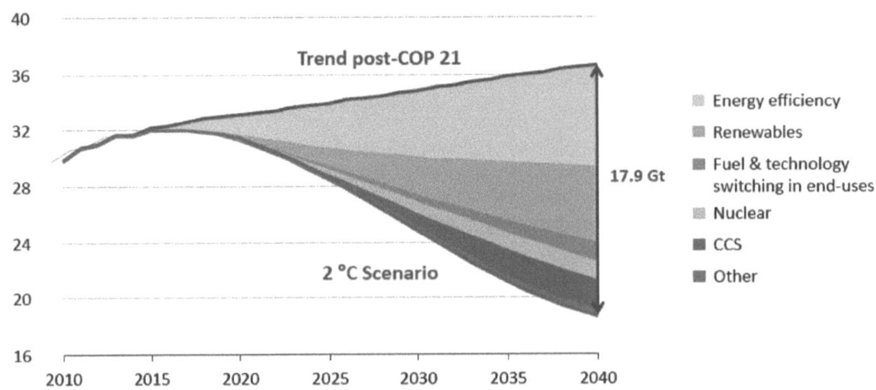

Figure 4.9. *CO_2 emissions after COP21 [IEA 16]. For a color version of the figure, see www.iste.co.uk/kayakutlu/energy.zip*

It is clearly seen in the IEA analysis shown in Figure 4.9 that clean energy deployment is behind, and therefore more support is to be given for research on electrical vehicles, nuclear safety, energy storage and renewable energy technologies.

One of the major impacts of COP21 is seen as the increasing support that will be given to local governance, mainly urban management.

The EU-subsidized project Multidimensional Impact of the Low-carbon European Strategy on Energy Security, and Socio-Economic Dimension up to 2050 (MILESECURE 2050) announced its policies as an output of the project in March 2015. These policies attracted more attention after COP21.

The agreement sets out a global action plan to put the world on track to avoid dangerous climate change by limiting global warming to well below 2°C. The agreement is due to enter into force in 2020:

– a long-term goal of keeping the increase in global average temperature to *well below 2°C* above pre-industrial levels;

– to aim to limit the increase to *1.5°C*, since this would significantly reduce risks and the impacts of climate change;

– on the need for *global emissions to peak as soon as possible*, recognizing that this will take longer for developing countries;

– to undertake *rapid reductions thereafter* in accordance with the best available science.

The project on energy transition is finalized with three main findings: social, political and grassroots factors are the foundation for an efficient energy transition; regional- and national-level bottom-up initiatives can only be scaled with external finance; behavioral change is the prerequisite for success of transition.

The report states that in order to leverage a transition political leadership, individual open-sightedness and human capital are as important factors as positive economic outcomes and market signals. Transition projects can be triggered by the engagement of individuals based on incentives and regulation but with attractive funding and disruptive outputs. New energy transition impacts can be summarized as new political power, new socio-cultural power, new business models and new economic power structures with new interactions of individuals with technologies [EUR 15].

More emphasis is given to financial models to engage all participating countries in reducing CO_2 emission decided by COP21.

The International Monetary Fund (IMF) reported a study in January 2016, which is the first report on microeconomic impacts of a greening meeting. This report also suggested financial policies imposed by the decisions taken in Paris [FAR 16b].

Impressed by COP21 decisions, June Sekera launched new economic approaches in public economy [SEK 16]. Again following COP21, the most well-known ecological economists came together in the book *Beyond*

Uneconomic Growth [FAR 16a], suggesting new financial approaches that give a smaller role to the IMF and a bigger role to sustainability.

In waiting for the actions following the commitments taken in Paris and their results, we mention some components and initiatives of "greening". The most important action is education and training. It helps evolve mentalities and professions as well as train new professions created by renewable energies.

4.2. Training for greening

The education of new attitudes respectful of our environment is a prelude to successful facing of challenges. Developed countries are used to abundance and waste. The new attitudes are radically different, that is why the education task is so tough. Moreover, renewable energies create new professions that need to be learned.

Applying the holonomy principle, if the culture of greening is spread to individuals and groups, businesses may pay more attention to their actions and give a greater importance to green products.

An important part of greening is renewable energies, which seem to grow to take 36% of the energy resources used globally. Even in 2030, heating seems the most effective application of renewable energies [SAY 15].

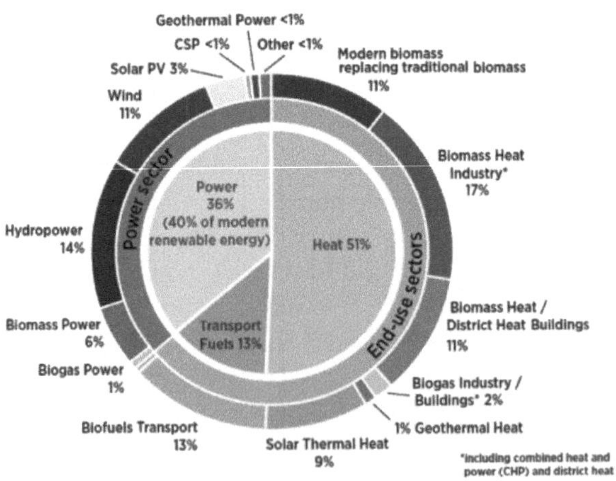

Figure 4.10. *Energy resource distribution with utilization field [SAY 15]*

The US Department of Energy stated in an early report that 70% of the cost of greening is on fossil energy and that the environmental concern had not started until 1985. It is a pioneering report in identifying greening triggers. In that report, the causes of greening are summarized as economic, environmental, security and knowledge benefits. It is obvious that investments in renewables have increased with the limits of fossil resources. It was almost looking for fuels in response to energy crisis, though majority of the research was given to making fossil energy more effective. This could be considered as the stage 1 of greening. Then came the stage 2; detailed measurement was possible for the environmental effects and the projection of harms on life was discovered. Fossil energy is marked as "dirty" with two million tons of SO_2 and 25 million tons of NOx emitted into air. The third stage was handling national security concerns, when enhanced oil production, field demonstrations, seismic and drilling, completion and stimulation projects were supported. Knowledge benefits were achieved as the most impressive result of all the technological research. New planning, optimization and management techniques were the new contributions to the business world.

Greening projects were also critical in the sense of partnership. The technologies, methods and global objectives enforced the partnership of the public and private sectors besides inter-industry collaboration. For example, advanced gas turbines were developed by an effective collaboration of chemical and paper and pulp industries with energy industries.

The majority of US or European research programs for the energy industry were more successful when coupled with an application program. Performance measures of the research programs should be made as a portfolio of benefits not only to the return on investments. When more than one benefit of the research is found, the multi-objective optimization of the performance becomes unavoidable.

Training on renewable energy is an important component of greening, but two complexities have to be considered in training programs:

– non-recognition of renewable energy as a principle component;

– multi-disciplinary and diverse nature of renewable resources.

Though there are several European or US universities giving graduate educations, there is no widely accepted program for the lay people who misuse information. Web-based training is provided by lots of private companies, but very little is done for education in secondary schools. Private organizations should pay for renewable energy programs for company employees.

It is evident that the current training should be focused on the job market. Early European training on solar and wind energy have not created the expected number of jobs mainly because of small starters in the field. The new focus is designed on regional requirements including rural areas and underdeveloped countries [US 14]. The cheaper the technologies, the higher the number of jobs. It should be noted that the number of technicians required are multiples of engineers requested in the renewable fields.

Educational documentation and reference manuals are an important part of this field. Case-based training of the World Health Organization (WHO) from the beginning of the century should also be a guide for rural electrification that has been targeted since 2014. The annual summer school for solar power organized by the UNESCO is a good example for these rural areas.

Continuous training in greening is needed as seminars, workshops and summer schools are targeting decision makers and experts related to other fields of energy and other industries. Different presentations for market penetration with the renewable energy progress with alternative scenarios are to be designed. Publications for wide access and audiovisuals given in the media will certainly add a great value of recognition. Regular broadcasting documentaries with visits to the renewable energy generation fields will be a way to avoid ignorance. A general store of basic training for different ages can be programmed. Web programs can be adaptive so that small changes in target audience (researchers, engineers, technicians, decision makers, industrialists, end-users, the general public, children, housewives, etc.) can be applied.

When the industries learn that greening is not only green features of the new products but a lifestyle to think of all the simple activities to be taken green cautious, it will be a wide success.

The power optimization expert Sterman defined the bridging of the systems to the greening of the year 2050, in 1982, as shown in Figure 4.11 [STE 82]. He emphasized a need for "backcasting" as well as forecasting.

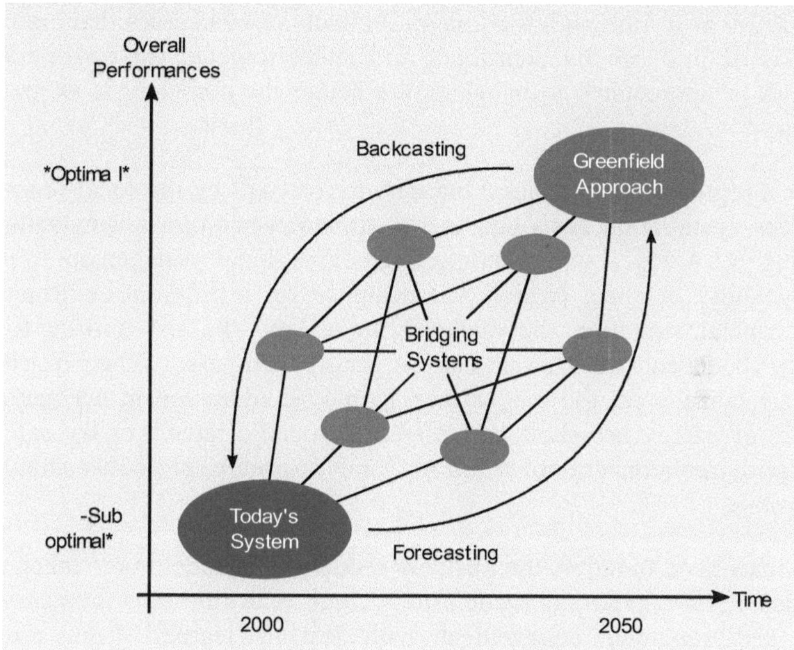

Figure 4.11. *Transition to greening [STE 82]*

An integrated multicarrier system used in bridges requires combined optimization to find the overall performance. The basic question in those bridges is to plan which energy carriers should be used for how much energy consumption in order to meet specific loads. This is a strategic question which could be answered in short-term operations, but has to be combined for long-term plans. Multi-objective optimization is to be performed to structure the composite objective functions based on minimizing costs and emissions while maximizing security.

Greening inspires the application of portfolio approaches because of multiple resources. Investors who tended to invest in hybridized renewable energies initially used this approach. But now, it has become a strategic framework for evaluating the impact of greening actions on the economic performance of a firm, industry or country.

From a corporate perspective, the impact of greening measures shows if the business is in line with the image. Products or businesses that are not as green as desired can be identified, and either targeted for environmental upgrades or approached to benchmark whether the company is as green as desired.

For a regional management, the key drivers will be the local renewable resources, installation costs and policy structures with an understanding of greening [VLA 15]. It will certainly force the regional management to renew the regulatory compact (which was designed for a different environment). The financial structure, the business model and the risk profile for the utilities all depend on the regional regulatory structures. When distributed users are considered, the inequalities should be addressed in net measures. Descent of energy use should be prized in specific rates. For the safety of local grid, interconnection reliability and operation protocols should be transparent.

Utilities need to adjust the business models to include more stakeholders and also renew operation models to include real-time operations, system plans and customer engagement with lessons learned from previous applications.

The European Local Energy Assistance (ELENA) program announced in 2016 helps the development of industrial companies with credits from EU treasury.

4.2.1. *Role of technology in training and education*

Energy and renewable energies offer a large spectrum of opportunities for new and traditional jobs. However, professional training should be attractive and available online and on mobile phones (e-learning and m-learning). While the field of diagnostics is well understood, training on how to compose an adequate offer with locally available resources is just emerging. Partners of the EU proposal ACREOS suggested a training that included a simulator for learning how to decide the best solution. The *Institut de Régulation et d'Automation* (IRA) among others, offers trainings in 3D immersion to allow trainees to learn through "real" situations [IRA 08].

Tutor systems for energy training working closely with experts have been used in France since the early 1980s [ROO 87]. Many others expert systems have been realized, covering all fields of energy production and use. The advantage of using the expert system for training is that it provides explanations of experts' reasoning in problem solving and decision-making.

While the above systems are still in use, the current trend is to put them online and design more serious games. EnerCities is a game open to all wanting to learn how to manage energy and sustainability [ENE 11]:

> "The EnerCities project has offered a game-based learning platform for young people to experience energy-related implications. The goal is to create and expand virtual cities dealing with issues such as pollution, energy shortages, and renewable energy. The European rollout of the platform – accompanied with education materials – was facilitated among 117 education institutions. Over 70,000 youngsters used the game-based platform. Also, six regional events, four game-competitions, one community platform, and one final conference were organized. The project made youngsters (and teachers) aware of energy-related implications and had the ability to change their energy-aware attitudes and behaviours. The platform was, and still is, attractive for youngsters: advanced on energy topics, teamwork-oriented, competition-based and community-driven. Educational institutions now have a learning tool to upgrade their education programmes".

The transition from traditional to renewable energies and in many cases a combination of both is an innovation *per se*. The field of energy, especially in the greening context, is an interesting playground for new experiments.

4.3. Innovation for greening

All the actors in energy and its related fields need to innovate, i.e. innovate in the way energy is produced and distributed, innovate in the production tools and processes, innovate in the way of managing distribution and influence the frugal consuming of energy by industrial, organizational actors and citizens.

All types of innovation are useful [MER 11] – in management and organization, in architecting Information Systems including modern methods and needs-oriented thinking instead of traditional blocks.

While technological innovation has always been an engine of progress, useful and friendly technology should be used instead of technology without "switching-off the brain".

The economic situation of many countries needs a social innovation in the manner that energy is optimized, used and shared. The awakening on the way of using the energy may come from social networks, serious games but mostly from shortages. In this last case, energy users should become smarter and use their knowledge to find the best solution or use multi-agent systems.

From the economic point of view, one of the greatest European challenges is growth and job creation. The field of energy has a potential to successfully face this challenge and obtain significant results.

4.3.1. *Innovation for economic development*

In 1999, the cost of increasing the central heating temperature by 1.0 degree was calculated as $5.5 million in the USA [ROS 95]. Today, it is estimated that the cost of this increases heating bills by 1 to 3%. With the increasing energy cost, more capital was put in energy-efficient technologies. However, with the current decrease in natural gas prices, there could be an increase in the demand. Moreover, it is known that companies take more energy conservation actions with the increase in costs. It has been analytically shown that rising energy prices are more effective in energy saving and cost emission reductions [LIN 12].

When minimizing cost was the only concern, new standards were defined in using fossil energies, intelligent and protective controllers were applied and modular converters were used considering the reliability considerations, and now the focus of research is on maintainability [BLA 05]. Furthermore, it is already known that today's mature technologies (hydro, wind, solar PV, biomass) will be more cost-competitive in 2030 [SAY 15].

The rate of economic development has been observed to have a correlation with energy consumption and production. The generic relation with electricity consumption is given with dependence on the electricity consumption, the GDP, the cost of electricity given for consumption and the total employment [TAN 12]. This definition shows the high interaction of the development and energy status of a country. A panel study conducted on OECD countries has shown that only Australia, New Zealand and Iceland have a very small impact on economic development from energy consumption, and that all the rest of the countries have an indispensable correlation [ASL 13]. For highly import-dependent countries like Turkey, development can be triggered by decreasing at least the industrial consumption prices [MUC 09].

Innovations can change the levels of energy dependence. Shale gas richness in the USA reduces the dependency of the USA on Middle Eastern oil. Renewable energy policies in Germany reduce their dependency on Russian Natural Gas and open the door for exportation. A study on four energy exporting countries based on fossil energies has shown that because of the trend for renewable energies, economic development is only influenced by the energy exportation amount rather than consumption or production [FEI 14].

The most recent IEA Outlook Special report suggests that investing in energy production and efficiency in Africa can fight poverty [US 14]. This will certainly change the energy use performances given by energy intensity. Energy intensity can still be expressed as the rate of consumption per GDP, but the energy cost intensity has to be taken into account as well.

Energy-related costs have a negative impact on the national wealth which is why the index should be recalculated. Radovanovic and Filipovic state that in three biggest energy consumers of Europe (UK, Germany and Italy), consumption has been reduced to less than 1 KWh per GDP unit [RAD 15]. This result also proves that developed countries invest more in research and innovation, which results in a better balance between energy and the economy.

The current energy crisis is defined by a dependency on foreign resources and failing to restructure the grid operations and energy trading market. An Australian study has shown that there is a considerable influence of the energy crisis on the economic crisis [CHE 13].

4.3.2. *Innovation in renewable energies*

The issue of innovation has always been the framework for technological progress in the field of energy. Innovation of new energy technologies is a broad set of knowledge consisting of new fuels, low-cost tools affecting the energy generation, as well as energy conversion.

The IEA defines three generations of renewable energy technologies [ENE 06]:

1) The first era of energies has reached maturity in terms of hydropower, biomass combustion and geothermal energy.

2) A second generation of rapid development in terms of solar energy, wind power and modern forms of bioenergy.

3) Most recent technologies presently in developmental stages in terms of concentrating solar power, ocean energy, improved geothermal and integrated bioenergy systems.

	2004	2014
Investment in renewable energies (billion US $)	45	270
Total power capacity of renewable energies	322.9	1712.2
Hydro-Power (GW)	36	1055
Bioenergy (GW)	227	93
Geothermal (GW)	8.9	12.8
Solar (PV) (GW)	2.6	177
Solar (thermal-CSP) (GW)	0.4	4.4
Wind (GW)	48	370
Solar hot water capacity (GWth)	86	406
Biodiesel (billion liters)	2.4	29.7

Table 4.2. *Renewable resources available [REN 15]*

The investment in renewable energies and current capacities are shown by the Renewable Energy Policy Network in Table 4.2 [REN 15]. It is clearly observed that the main investment is given to the field of renewable energies.

Making new technologies economically viable affects both the demand and the supply of energy. Technological innovation continues supporting wider utilization of renewable resources. Innovation in renewable technologies can be summarized by efforts focused on:

– decreasing the weight and sound of wind turbines;

– increasing the power generation efficiency of wind turbines;

– reducing the cost of solar energy while increasing the efficiency;

– using bigger variety of plants for replacing food plants for biofuels;

– increasing the potential of heat and power from biomass with lower cost;

– increasing the ways of using wave and tidal energy.

The use of renewable energies for more reliable heating that can reach high temperatures quickly, like plasma techniques, has been part of innovation in heating. In low-income regions, it is valuable to use biomass for heating both buildings and water systems. Concentrated solar collectors have a wide potential of applications in heating. The largest diffusion of the renewable energy use in heating is expected by the use of heat pumps and different technologies of the pallets.

Improvements in heating industries have expanded to cooling as well. The tri-generation, which is combined heating power and cooling, has been one of the greatest innovations in the general use of renewable energies. District heating has been well implemented by using the renewables in Europe. The building heating systems has been renovated in many countries including the USA. Legal allowances for micro-generation have encouraged the industrial use of renewables for the combined purposes of heating, power and cooling. The availability of heat storage and more power storage systems in a number of European countries has strengthenend the district heating approach. The Renewable Energy Policy Network reports on the use of renewables for the combined energy generation are used in district heating by at least 20 European countries and for industrial purposes in 105 countries from Chile to India to the United Arab Emirates [REN 13]. District heating based on the use of renewable energy has supported local grids. The renewable report states that "in 2013, China called on high-wind provinces to begin pilot testing of wind-to-heat technologies".

There is a continuous discussion that the production of energy-efficient products and systems may lead to lower energy consumption. It is suggested that national efforts for the green actions on the macro-scale can lead to continuous value creation. The research conducted by Herring *et al.* shows that only 10–20% of technical innovations cause a reduction of consumption in the short term and contribute 1–2% to the income increase [REN 13].

In addition to the intelligent equipment and utilities used in power generation and distribution, the main innovation was the intelligent control of energy consumption in residential office buildings. El-TERTIARY is a European Project that analyzes the typical use of intelligent control. This project studies schools, hospitals, offices and all other buildings based on energy use [GRU 08].

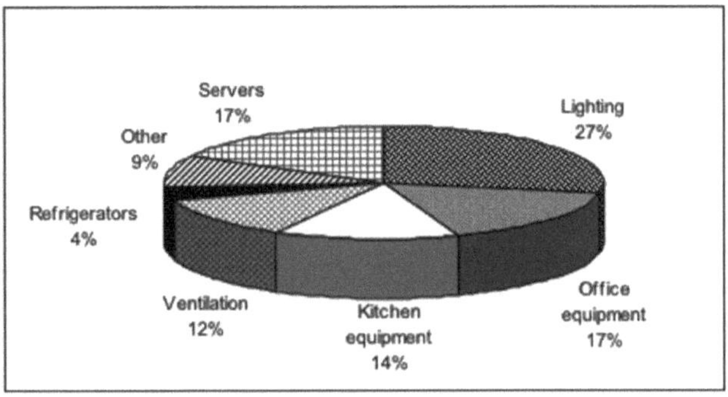

Figure 4.12. *Distribution of energy use in office buildings [KRA 10]*

According to energy engineers, intelligent control in a building aims to optimize energy use through control circuits that are installed in thermostats, dampers and rooftops. Environmental heat transfer is controlled in insulated and/or non-insulated buildings by considering the different scenarios of a thermal network to optimize the inner temperature, environmental temperature and the heat transfer coefficient [KRA 10]. Intelligence in buildings is not limited to control, but wireless sensor networks and/or outdoor nano-materials are also used to sense the environmental temperature with the aim of creating thermal comfort indoors.

One of the widely accepted innovations in the energy field is fuel cells. They are used with great excitement mainly because electricity generation with fuel cells has the highest efficiency starting with 35% and reaching up to 55%. Moreover, because they do not have any moving part during the power generation, they have high environmental-friendly features. Fuel cells are produced in the form of pure hydrogen, phosphoric acid, carbonate and, most recently, solid oxide. Electric load using fuel cells and batteries together will respond the thermal needs. Low-temperature fuel cells are currently researched to be made available for vehicles [MOZ 07].

4.3.3. *Innovation for development of social values*

Energy-efficiency and ecological concerns can be gained by a society following social education efforts and a considerable paradigm shift in the energy-based culture. This culture will lead to renewable energy preferences. The UK is one of the biggest consumers in Europe, where a people-centered eco-design project research was realized in 2006 by linking the surveys with a climate change TV program on BBC [HER 07]. Reasons for changing to loft insulation, photovoltaic heating or micro-generation technologies were stated as cost reduction, due to ecological concerns and for warmer houses.

Though capacity limits of fossil energy triggered the ecological consciousness, improvement in energy cognition is only possible by considering it as a social value. Energy is a source of knowledge that needs attention based on daily purchases (as the block marketing in different daily sections are possible) to the political choices. Energy regimes and even democracy are related to the use of different energy resources. Mitchell has put forward the theory that carbon usage with local economy transferred to oil democracy of internationalization has already triggered technological developments. Followers of Mitchell [MIT 09] have observed high innovations in energy resources and technologies. However, energy technologies are also seen as part of the cause for degeneration in improved democracy [AGU 12]. Social sciences are necessary to help society accept the social values of energy. As Spreng stated, engineers, economists and social scientists need to work together for the future of energy [SPR 12].

As Dias [DIA 06] demonstrated at the beginning of the 21st Century, country development levels are not in line with the increase of energy usage efficiency, but human development shows a greater need for energy conservation education.

4.3.4. *Innovations in energy governance*

Energy governance involves the processes of planning, scheduling, designing, implementating and monitoring rules and agreements related to energy. Due to the liberalization of the utility sector all over the world, the actors of the energy industry are concerned. Governments, non-governmental organizations, civil society groups, corporations, citizens, public–private partnerships and ordinary consumers are part of governance. The roles differ in scale, from the individual household to the local region or to the national and global level.

There are standards defined to make the management of energy easier. Building standards in the UK are based on heating, considering the improved boiler efficiencies and reduction of losses in hot water transfer arranging the temperature to handle the weather changes. Europe has defined the standards for energy networks according to centrality, country GDP and ease of adoption to the rules [MAG 14].

The USA, the UK and Russia are the big players of the oil industry. As energy resources shift toward renewables, Germany and China have also taken their place among the biggest players. The UN has made attempts for global governance. However, in 2008, the International Partnership for Energy Efficiency Cooperation (IPEEC) took a big step in global governance, followed by the International Renewable Energy Agency (IRENA) in 2009 [KAR 10].

The majority of these countries prefer to join international governmental organizations, believing that even after several levels of privatization, the energy governance will be a major security concern. Baccini showed that although producer countries prefer partnership with the competitors to follow the market closely, the reverse is not correct for the consumer countries [BAC 13]. Furthermore, regional security complexity makes countries take shelter under the regional political umbrella. The designs of cooperation in international governmental organizations are based on energy trade choices and the geopolitics of the energy producing countries make up the organization.

Governance of renewable energies was designed around subsidies; this explains why renewable producers cannot take seats in oil and gas lobbies. There is now a need to enlarge the scale of renewable governance to include market support and operation planning. A recent analysis of further scaling of renewable governance shows that operational planning and market support will bring financially competitive advantages to both the renewable energy producers and the countries for leading the future energy pathways [COW 15].

The improvement of wind energy capacity, shown in Figure 4.13, demonstrates that there was a jump in 2009 when the smaller power reductions were encouraged in many countries with micro-grids and micro-generations.

Reliability and performance increases while investments are reduced by low power applications. Even modularization for bigger requirements and system integration for the medium requirements will soon be preferred. Fast energy storage will also be a future trend for thermal energy as well as power generations.

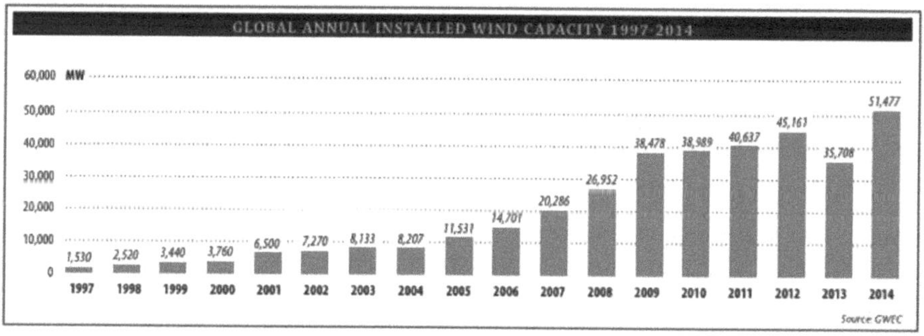

Figure 4.13. *Wind capacity improvement in the world [GWE 14]*

4.3.5. *Big data paradox*

As in any other industry, producing and sharing with stakeholders the on-line information, online training, internal social networks, e-commerce, advertisement and even recent retail power sales online generate a huge amount of multimedia data stored in data centers.

The production of data is "encouraged" by social networks, shared media and other Web 2.0 services. Cities, hospitals and other organizations also generate a lot of data.

Big data is considered as a potential source of revenues from services based on exploring data using analytics and other techniques.

The paradox is that datacenters need power to generate the heat that is eliminated by cooling systems that also consume power. Apple, Google, Cisco and Amazon power their operations with solar energy recovered from datacenters; they can also supply heating and hot water for houses, shops and public facilities.

So far, this energy is simply evacuated into the air by means of different air conditioning or water cooling systems. However, in recent years, recovery and reuse initiatives of these energies are in place to heat homes, offices, factories, swimming pools or even vegetable greenhouses.

On the University of Amsterdam campus in the Netherlands, there is access to free hot water from the Equinix datacenter. In Roubaix (north of France), the five OVH datacenters heat the company's offices and some neighboring firms. The datacenter of the British operator Global Switch powers a tropical flower greenhouse in Clichy, near Paris. In Uitikon, Switzerland, a public swimming pool is heated by an IBM datacenter.

The most impressive experiment in Europe is conducted in the business cluster Val d'Europe, near Disneyland Paris. The heat produced by the Natixis bank datacenter supplies a district heating network. Hot water from the condensers of air conditioning systems is recovered by heat exchangers, and then distributed over the network using a central energy production. Ultimately, this system circulates water at 55°C in a network of 4 km of pipes, connected to a variety of buildings.

Innovation in the way of preserving useful knowledge involves switching from data thinking to knowledge thinking [MER 11]. Knowledge can be acquired by computers using multimodal interfaces or directly M2M and which are stored as reusable blocs (like Lego). Such storing needs less resources and less power; semantic and knowledge model-based navigation is also effective, providing only relevant information for the user.

The greatest innovation will be a new business model in replacement of the current advertisement-based one. It will dramatically reduce the amount of data generated every minute.

4.3.6. *Creativity for innovation*

The biggest challenge for innovation in energy (and other fields) is flexible and creative thinking [DE 80, BUZ 03, MER 11].

Charles Savage set up the foundation of Industry 4.0 in his book *Fifth Generation* published in 1996 [SAV 96]. In the book, he writes about the knowledge era as a pioneer and gives a new dimension for dynamic and shared optimization, which cannot be avoided in the energy industry. The first bricks of knowledge era signaled for the growing importance of optimization. Currently, he teaches students to deal with multiple crises.

One of his maps is shown in Figure 4.14. Savage emphasizes the capacity of thinking global and dealing with any consequences simultaneously [SAV 10].

Figure 4.14. *A sample education map from Charles Savage*

Creativity, capacity to apply the various reasoning models, making links and using the available knowledge and right technology are *sine qua non* conditions for sustainable innovation. A sustainable future requires more respect of our natural ecosystems, among others. If individuals, organizations, companies and politics are respectful, we will be able to achieve the COP21 goals through greening the new technologies, new resources and new applications in the energy field.

The next chapter presents some new trends, experiments and promising perspectives for sustainable future of energy.

5

Future of Energy

5.1. Paradox of technology and human

Many factors influence the integration of technology into all human activities. The accelerated development of devices and software pushes the energy market to try and apply various technologies for optimizing, follow-up, switching and remote control. The context of smart, innovative, green and wise cities, innovative buildings systems, city farming, “greening” of datacenters and many others is based on technology.

While the enchantment by technology is growing and technology is strongly pushed to all fields and activities, paradoxically, technology should be powered and hence produces a lot of waste. We change tablets, computers and smartphones very often and the new generation uses more energy because it provides more “services” that we do not necessarily need. The users of social networks produce a huge amount of data stored in powerful data centers that need to be powered and cooled.

Technology has also a cognitive impact – the more we are assisted by technology, the less we think. In many cases, technology is not necessary to perform a given task, but it is here and we use it mechanically without searching for alternative solutions.

At home, we have a lot of various devices that are permanently powered or in standby. After several televisions, microwaves, kitchen aids, laptops, play stations, fridges, freezers, air conditioning and home automations, now connected objects and robots make their entry.

Many of these devices can be replaced by zero energy similar activities; however, it requires an alternative thinking and sometimes applying the solutions from the past. For example, in summer, many cooling devices are automatically switched on, while we can close windows and use stores, or apply natural regulation of temperature using adapted materials, or combination of the two. A traditional provincial house has thick stone walls, small windows and porch roofs. The shutters are closed to the hasp to avoid the entry of hot air. The sleeping rooms are north oriented; spaces of the trees are carefully selected to grow in the dry soil and provide shade.

Figure 5.1. *A building in New York and a house near Montpellier, France*

We still do not know the function of "air shafts" in Egyptian pyramids. Were they conceived for cooling inside?

Today, air conditioning in buildings and in cars is overused, the same for the fuel, because many people do not practice the intelligent use of energy and economic driving. Being continuously supported by technology, they forget to think!

The use of energy became mechanical, and few consider that our resources are not infinite.

In the current context, the education of new (or old) ways of doing and appropriate attitudes is necessary. Today, education can use serious games such as efficient use of water [EMP 16] (http://www.empreinteh2o.com/) or energy [INT 13] (http://www.2020energy.eu/en). But sometimes discussion with elders in their nursing homes is beneficial for all.

A new attitude should be developed – think before, use your own or collective knowledge and find the best solution.

While small companies make savings in using energy, large companies and other organizations such as universities, governmental organizations and NGOs should also awake to a smart use of energy. The main problem is that the employees or civil servants are not charged for the use of energy and, in consequence, are little motivated for its intelligent use.

Applying a holistic approach to energy management may help influence the working environments – if an individual is educated in smart use of energy, he/she will apply the same attitudes in their working/living environments.

Most activities in all fields are now supported by information technology. The information systems should be powered. Technology designers need to make an effort in intelligent and adaptive programming as well as in hardware and software greening.

The software programmer should also "go green". There are still many unused programs in our laptops and other devices consuming energy unnecessarily. Intelligent programming, introduced by artificial intelligence, allows building the reusable models of knowledge using KADS, ontologies, cases and others [MER 11]. Fitness for software, operating systems and applications is more than suitable.

Considering the huge amount of data produced worldwide, efforts are made to find the most effective management of energy for powering and cooling datacenters [MER 15].

The new approaches to eco-design [FEN 13] progressively replace the traditional methods. Eco-design considers the whole lifecycle of a given product, including recycling. Simulators help evaluate the impact, choose the right materials, and evaluate the energy and water needed throughout the production process. The eco-innovation process that also applies to eco-design is shown in Figure 5.2.

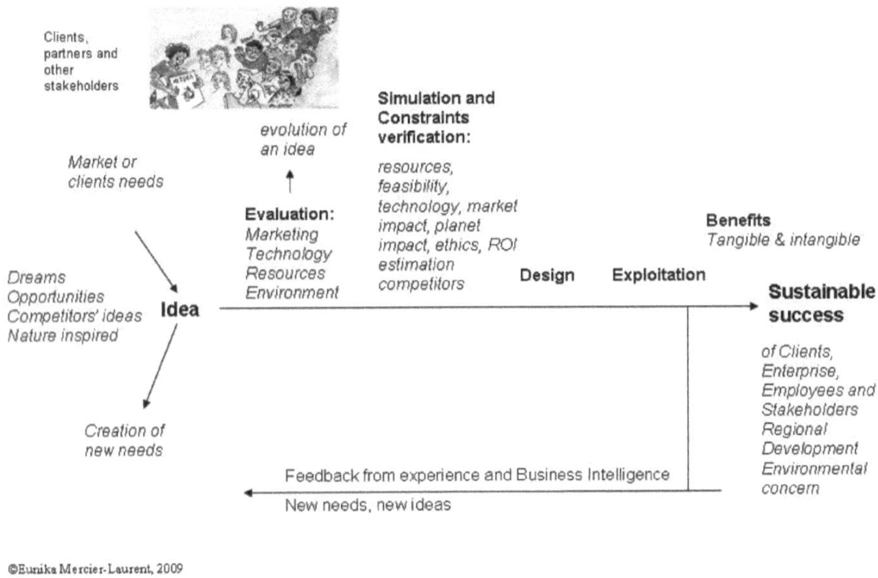

Figure 5.2. *Eco-innovation process [MER 11]*

The eco-innovation process is composed of two stages: creativity and transformation. The ideas may come from inside or outside of a given organization. Two ways are possible: address the existing needs or create new needs (disruptive innovation). The ideas expressed by participants or generated by "electronic idea generator" are evaluated and the best ones are selected for transformation. Prior to design, the verification of all constraints such as competitors, market impact, technological feasibility, return on investment, impact on the planet and living, availability of raw materials, recycling, etc. should be made. These verifications, which may use simulators, will help decide the best choices for functions, materials, shapes, interfaces and business models. Sustainable success is guaranteed with the quality of those verifications.

The intensive development and use of technology has multiple impacts, which should be considered. While corporate social responsibility tries to balance the environmental impacts on health, society and economy, cultural, technological, cognitive and political impacts are not considered in current activities.

Almost all activities are now supported by ubiquitous technology. While technology provides considerable help, sometimes it also creates risks. Machine-to-machine communication prevents humans from decision-making, and connected objects suggest solutions that we are able to find without this help. Is it a sustainable solution? The added value of humans is still its knowledge and experience.

The recent approach is that the industrial renewal tries to face the economic crisis by adopting technology everywhere. Big data, cloud computing, connected objects, machine-to-machine communication, 3D printers, drones and robots change the industry but also need to be powered.

What is the right technology? Mostly, the traditional integer programming model is applied, because it is known and used by many. New technologies are much more powerful but require thinking differently; there is a need for switching from traditional IP (database, trees, etc.) to complex system thinking and considering the various impacts. These impacts are not only environmental, economic and social, but also cultural, political and cognitive and may inspire new technologies conceived in synergy with our natural ecosystems.

The artificial intelligence approaches and techniques have been developed to really help people in their activities [RAU 85, MER 07, EUR 16a, EUR 16b, EUR 16c].

For several years, natural intelligences have been observed and copied – plants, insects, birds, trees – and we try to understand how they produce the energy for their activities and how this can be reproduced for humanity without destroying our natural ecosystems. For example, birds or butterflies are able to fly long distances without stopping [MER 15].

"To keep themselves warm during cold weather, butterflies spread their wings in the Sun. The wings of some species of swallowtail are remarkably efficient at trapping and absorbing sunlight. The insects' secret lies not just in their dark pigment but also in the structure of microscopic, overlapping scales coating their wings. The scales, in turn, contain rows of honeycomb like holes separated by inverse V-shaped ridges that funnel light into the holes. This ingenious structure traps incoming sunlight, making the wings extremely black and warming the butterfly with amazing efficiency.

Butterfly wings may rank among the most delicate structures in nature, but they have given researchers powerful inspiration for new technology that doubles the production of hydrogen gas – a green fuel of the future – from water and sunlight" [SCI 12]. Other promising applications include optical instruments and solar cells.

Reducing the energy required for heating is a matter of life and death for penguins, and being warm-blooded means that they have a hard time keeping their body temperatures high. Part of the solution is an ingenious heat exchange system in their feet. As the warm blood reaches the extremities, it passes the cold blood traveling in the opposite direction, back into the body. This process cools the blood entering the feet and, therefore, reduces heat loss. This technique may be used to reduce energy use in many building systems [MER 15].

Plants are able to generate energy. They are complex organisms that rely on the transport of ionic signals and hormones to perform everyday functions. Controlling the chemical pathways in plants could pave the way to photosynthesis-based fuel cells and devices that modulate the internal functions of plants so that they produce important molecules more quickly, such as those needed for medicines.

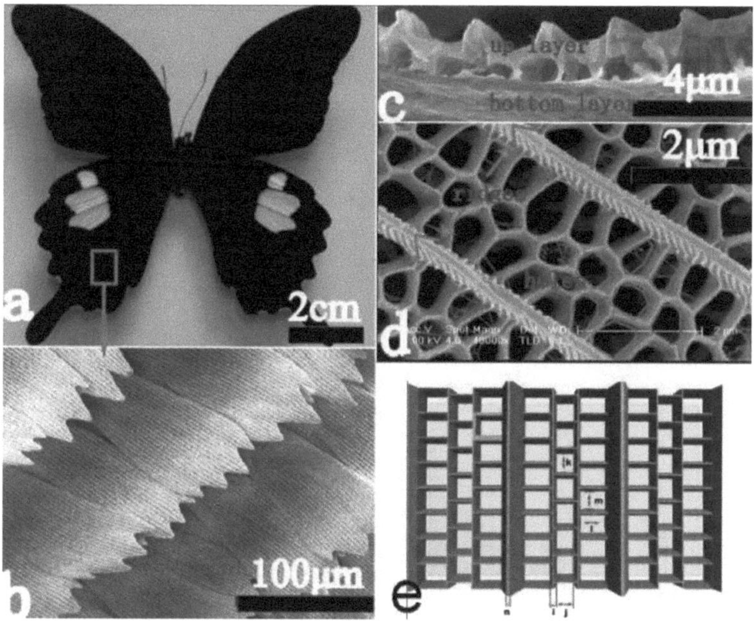

Figure 5.3. *Architecture of butterfly wings (source: American Chemical Society)*

A team at Linköping University in Sweden 'wired up' a garden rose by setting the flower into a basin of water containing a soluble polymer that conducts electricity.

As the rose sucked up the water, it also took in the electronic material which then integrated into the flower's own biology. It is the same process that allows flowers to be colored by dipping the stems in dyed water.

Once inside the plant, the polymer was designed to self-organize to form a wire with the help of the ions that were already present in the flower. The team showed it was possible to create an electric circuit using the embedded wire.

"Now we can really start talking about 'power plants' – we can place sensors in plants and use the energy formed in the chlorophyll, produce green antennas, or produce new materials. Everything occurs naturally, and we use the plants' own very advanced, unique systems." [KNA 15, STA 15].

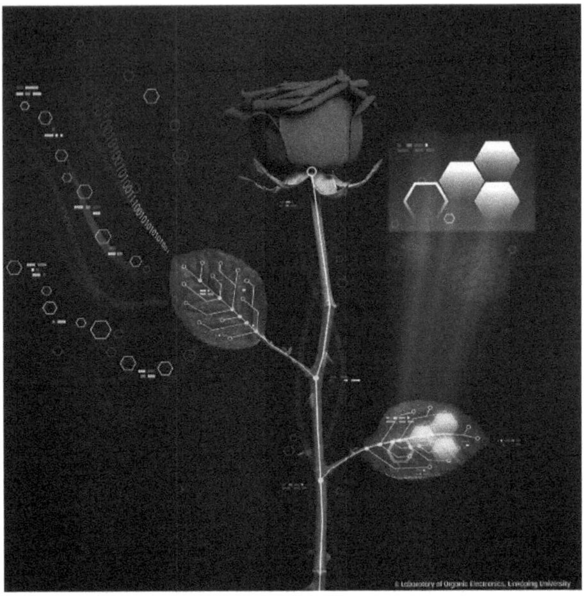

Figure 5.4. *Rose communication system*

Other researchers have worked on the adaptive capacity of plants, their resistance to metals [FED 14], and the capacity of absorbing radioactivity [SHT 04] or electromagnetic radiation or purifying air.

Artificial swarm intelligence was inspired by ants and bees [PEN 06]. These techniques are currently used for optimization and decision support systems.

Technology and, in particular, computers in all forms have to provide intelligent support to humans, performing tasks in which they are better than humans, but not replace them.

The storage of renewable energy is among seven challenges for France, presented in the report by Anne Lauvergeon, the former CEO of Areva [LAU 12].

Actually, the new wave of industrial renewal is among the challenges for many developed countries. This renewal is mostly based on technology, but has ambitions to generate jobs and growth while preserving our planet.

Recently, China has become the factory of the world; so what is the role of other countries? How will worldwide logistics starting from China affect energy use and climate change?

It is very critical to ask how renewal strategies will influence smart energy use. Which technologies are essential to support both growth and sustainable development?

5.2. Industrial renewal

Industries are still huge consumers of energy, but it may change with industrial renewal. Globalization and crisis spawned by it in developed countries have influenced technological progress in terms of digitalization, miniaturization and integration. This progress, as well as the existing approaches and technologies, are the pillars of Industry 4.0. The challenge is to transform production from isolated, optimized cells to a fully integrated, automated and optimized production flow, leading to greater efficiencies. It is supposed to change traditional production relationships among suppliers, producers and customers – as well as between the human and the machine.

The Fourth Industrial Revolution is triggered by the massive introduction of technologies such as sensors, the Internet of things, 3D printers, cyber-physical systems (CPS), clouds and automation into production systems.

5.2.1. *Technological components*

An abundance of existing technology is available. While all are strongly pushed to the market, we need to know how to choose one or a combination of several suited for a given activity or problem solving.

5.2.1.1. *Internet of things (IOT)*

Since the beginning of expert systems, sensors have been used for automatic data collection along industrial processes, but only some of them are networked with machines to make use of embedded computing. Collected data feed decision support systems [DOL 96] for process control. Today, sensors are miniaturized and connected.

With industrial IoT, more devices are enriched with embedded computing and connected using standard technologies. It allows field devices to communicate and interact both with one another and with more centralized controllers, as necessary. It also decentralizes analytics and decision-making, enabling real-time responses. Products are identified by radio-frequency identification codes, and workstations "know" which manufacturing steps must be performed for each product and can adapt to perform a specific operation.

5.2.1.2. *3D printers*

The invention of the 3D printer introduced a revolution in the way that product prototypes, and even ready products, are prepared. It is certainly less energy consuming to print a given component of a model, rather than following the traditional way of manufacturing.

The massive entry of digital products has taken off following 3D printers, and digital factories are approaching. The next stage will follow the virtual products (which are currently available) and virtual factories. In France, there are FabLabs for cheap and easy prototyping of new products offered to shopping mall companies. The greatest challenge of industrial renewal is to consider all the impacts of innovation for sustainable success. These impacts follow system logic and influence each other.

We expect that 3D printers will allow a more sustainable way not only for new product generation, but also for using the existing products. Today, we usually buy an average quality product conceived with a planned obsolescence strategy to push people to buy more. 3D printers may be used as smart local services for printing spare parts instead of replacing the product, recycling the old.

5.2.1.3. *Autonomous robots*

Manufacturers in many industries have been using robots to tackle complex assignments. They become more autonomous, flexible and cooperative, able to interact with one another and work with humans and learn from them. They are cheaper and offer more functions. Kuka and ABB offer such industrial robots in Europe, among others.

5.2.1.4. *Drones*

In industry, these small flying machines are used for taking pictures in inaccessible places, for survey and diagnostic purposes.

EDF [EDF 15], the national French provider of energy, uses drones for control and monitoring of the distribution network. Drones are also used for the detection of incidents and the monitoring of production facilities in the field of civil engineering [BOU 14].

Large companies like Google, Boeing and Northrup Grumman all have significant drone businesses because drones from the size of a palm to big plane are being used in the energy industry. One of the most widely accepted areas is to control the operations. Pure-play small drones developed by the US company AeroVironment have already traded publicly to be used in gas and oil sites to replace human inspectors who are more costly.

Industrial surveillance is the first field of industrial drones. Center dot data and image collection with geographical positioning system (GPS) are widely used. GPS is used in threat detecting or redirecting (like birds). Center dot images facilitate technology optimization and find the shortest path for tested applications.

Another very useful area is data collection by drones. If data is collected at the highest position a drone can fly, uncertainties in wind strength and solar illumination are reduced. Soil data can be collected at lower heights. There are drones that descend to collect power received from traffic and use the wireless energy mode without a battery when flying at a higher altitude [CAM 14].

Very recent studies have targeted the use of industrial drones in dynamic adaption of the energy systems. There are approaches like remote setting of the energy storage system according to the data collected by the sensors placed in a drone. Mobile energy storage becomes more efficient by using drones.

Drone utilization is indispensable for the management of microgrids as smart measuring is the most important characteristic. Daily multiple flights of the drone allow recording of heat and humidity changes at very small time intervals. Advanced prediction of precipitation allows balancing energy in the microgrid.

It remains to be seen whether or not drones could be the most powerful help in predicting the failure of nuclear energy centers. If we could track the security of the nuclear systems, we could certainly generate more clean energy. Yet today, we are still scared that the unknown drones get closer to the centers.

5.2.1.5. *Simulation*

Simulation has been used for years in 3D for products and production processes or for choosing the right materials in design; now it enters into the whole lifecycle of the innovation and eco-innovation process. In the future, simulations will be extended to plant operations. Then, it will include machines, products and employees to allow operators to test and optimize the machine settings virtually for the next product in line, before the physical change is applied. Therefore, machine setup times will be lower and quality will be increased.

5.2.1.6. *Big data*

Various activities have been contributing to the accumulation of phenomenal amounts of data. The intelligent access and exploration of this data has the potential to create new businesses. The techniques of knowledge discovery from data, text and images may help in intelligent exploration.

5.2.1.7. *Cloud computing*

This is a technology pushed by IT vendor companies, who are already using cloud-based software for some applications. With Industry 4.0, production-related data sharing across sites and company boundaries will increase. As the performance of cloud technologies improves, machine data and functionality will increasingly be deployed to the cloud, enabling more data-driven services for production systems. Even the control and monitoring systems may become cloud based. However, companies have to be careful about strategic data, information and knowledge shared on clouds.

5.2.1.8. *Cybersecurity*

For security reasons, production systems in many companies are not connected or closed. With the increased connectivity and the use of standard communication protocols, the need to protect critical industrial systems and manufacturing lines from cybersecurity threats increases dramatically. Secure, reliable communications as well as sophisticated access management of machines and users are strategic.

5.2.1.9. *Additive manufacturing*

Companies have just begun to adopt additive manufacturing, such as 3D printing, which is used mostly to prototype and produce individual components. With Industry 4.0, these additive-manufacturing methods will be widely used to produce small batches of customized products that offer construction advantages, such as complex and lightweight designs. High-performance, decentralized additive manufacturing systems will reduce transport distances and stock on hand.

For instance, aerospace companies are already using additive manufacturing to apply new designs that reduce aircraft weight and lower their expenses for raw materials such as titanium.

5.2.1.10. *Augmented reality and 3D immersion*

Augmented reality-based systems support a variety of services, such as selecting parts in a warehouse and sending repair instructions over mobile devices. Companies are making much broader use of augmented reality to improve decision-making and work procedures. Workers may receive repair instructions through augmented reality glasses. It can also be used for virtual training. Siemens has developed a virtual plant operator training module using a realistic 3D environment with augmented reality glasses to train plant personnel to handle emergencies. In this virtual world, operators can learn to interact with machines using their 3D representations [LES 16].

5.2.1.11. *Serious games*

From a practical point of view, the use of serious games for training is much more attractive and effective than the traditional power point presentation boring training sessions. Design of a serious game requires new capacities such as game designer and cannot be performed by a traditional IT engineer.

Serious games can be used for awakening [IEE 13] and for professional training as well.

There is growing evidence that some people respond better to quick information feedback, and this is why serious gaming has taken off in the business world. Using digital games for learning, simulation and demonstration has been observed to improve learning efficiency through increased motivation and relevant scenarios. Reports have shown that the players can be grouped as software experts, investors, intermediate players and targeted industries. That is why serious games can be an alternative learning method for energy generation and use. Though current games are structured based on the existing value chains, through the scenarios played, even new value chains can be established.

Since serious games define the goals, defining the learning and communication roles ahead, they can be used in teaching at any age, analyzing the market insight, planning and designing strategies. Since European strategies consider that efficiency in residential energy use is critical, in Europe, consumers are the first target group. However, in terms of resource use, government authorities can also be educated.

There is an unavoidable need to inform, inspire and educate young people about sustainability, energy consumption and conservation. Grossberg *et al.* [GRO 15] studied 53 games developed in 2010–2015, that aim to influence behaviors for energy efficiency and sustainability. This study highlighted the fact that games provide features that wire human brains to enjoy the challenges, social bonding and positive feedback. The evidence suggests that these games have the power to encourage a positive behavior change. They covered a wide range of solutions such as recycling, alternative energy, energy saving, utility efficiency, etc. between his/her neighborhoods; the ones categorized as complex included real-time energy use data. The featured characteristics of these games are observed as:

– rewarding when energy is saved as a result of some actions;

– competition to improve energy efficiency individually or in teams;

– using virtual worlds extensively;

– giving instant feedback to players on their energy use with real-time granular data.

The serious games developed on the initiative of association, Du flocon à la vague (from flake to wave) [DU 15] and Empreinte H_2O (water footprint) [EMP 16], are good examples of giving the desire to save and sustain water resources through games. The first combines the various sport challenges and related knowledge in physics, chemistry and environment. Players have to succeed in sport and knowledge challenges to win the game. Four seasons are represented to cover all sports having water as support (liquid, snow and ice). The second game allows evaluating the amount of water used for given activities. It also works as simulation – if the fish is dead, it means that the water is wasted and the player has to do better.

5.2.1.12. *Horizontal and vertical system integration*

As many companies are not practicing holistic knowledge management and their knowledge flow is partial or non-existent, most IT systems are not fully integrated and the communication between departments, such as engineering and production, and service, suppliers and customers is not organized. Functions from the enterprise to the shop floor level are not fully integrated. Engineering from products to plants to automation lacks complete integration. We hope that Industry 4.0, powered by knowledge management, will allow companies, departments, functions and capabilities to become more cohesive. Dassault Systèmes and BoostAeroSpace launched a collaboration platform called AirDesign for the European aerospace and defense industry. This platform serves as a common workspace for design and manufacturing collaboration, and is available as a service on a private cloud. It supports the exchanging product and production data among multiple partners. Before them, Airbus has already practiced it through a concurrent engineering concept.

5.2.1.13. *Cyber-physical systems*

CPSs are networked and computer monitored. They interact with humans and/or with other physical systems. Smart grids, autonomous automobile systems, robotic surgery, health monitoring, process control, distributed robotics and automatic pilot avionics are some examples of such systems. Designing CPSs requires transdisciplinary approaches including cybernetics, mechatronics, artificial intelligence, design methods and knowledge of the application fields. According to the function of applications, CPSs are adaptable, precise, autonomous, efficient, able to avoid collisions, able to operate in dangerous or inaccessible environments (nuclear plant, deep sea) and of multifactor coordination (traffic control, process control). Mobile CPSs include mobile robotics and electronics transported by humans or animals. They are used to track and analyze CO_2 emissions or other pollution, detect traffic accidents, evaluate traffic, provide help in case of accident and monitor patients. Security (prevention of terrorist attacks, of power grids) and cybersecurity remain the most ambitious challenges for CPSs.

According to [LEE 15], CPSs are part of larger five-level systems including control, decision support systems, prognostics and condition-based monitoring. This approach is shown in Figure 4.5. At the first level, a smart connection is required, where data is collected from all the machines used in the plant. The measurement of the data can be done by sensors or directly collected from the enterprise information systems. The second level is the conversion of data into usable information, in a way "giving self-awareness" to the machines. Calculations are done in this process. The cyber level acts as a hub of data collected from the network of machines, where the historical data can also be used to compare and predict performances. The cognition level requests a thorough knowledge of the monitored system. Presentation of all collected information should be prepared for the experts. Both individual data and comparative information are used to present the knowledge that will support the decisions to be taken. The last level of configuration is the level where cyber systems are converted into self-adaptive, self-configuring and self-repairing physical systems. Corrective and preventive decisions are taken at this level.

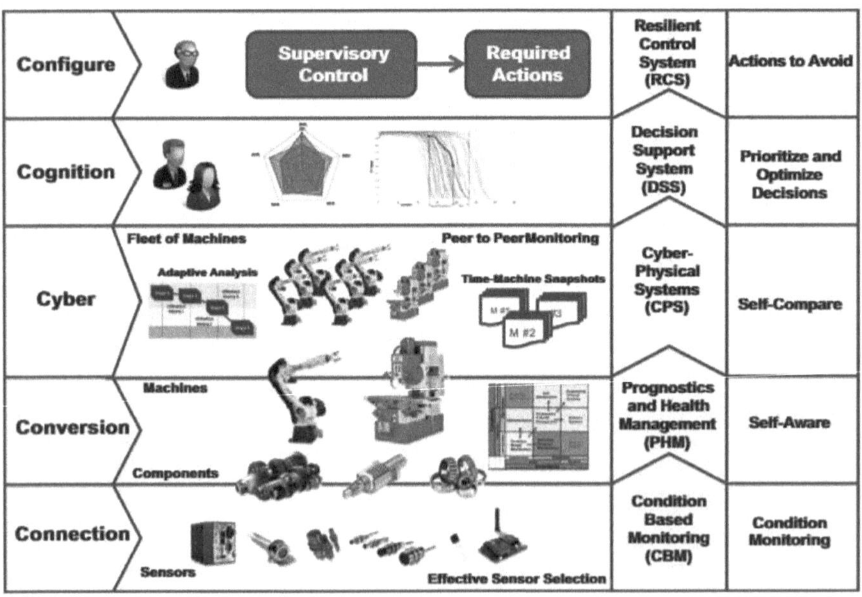

Figure 5.5. *Five levels of CPS according to [LEE 15]. For a color version of the figure, see www.iste.co.uk/kayakutlu/energy.zip*

Smart CPSs are considered as strategic for Europe and are among the EU H2020 topics (ICT challenge).

Flexibility is more about internal changes of the manufacturing companies as in capacity flexibility or material flexibility. That is why all the changes with Industry 4.0 are named as transformation. The need for transformation is in marketing strategies and competitiveness definitions that will influence marketing. Accordingly, the key indicators will be redefined and cause the internal changes.

Transformation to Industry 4.0 needs solutions for the following issues at the operational level:

Industrial machines: the similar or identical process machines are used in totally different conditions. There is knowledge to be gathered about those machines in different situations, which could help to standardize the ideal environment for the most efficient energy use in addition to the most efficient productivity.

Quality of products and processes: integrated software is required so that the quality department gives feedback to the production department for the health of marketable outcomes.

After the inclusion of applications in production planning and management, manipulating big data has become an issue. To explore and keep the data accumulated under control, self-learning machines are of interest. The main challenge is that learning should be not only from accumulated data, but also through the interactions with the users.

Human interactions: machines are run and managed by human operators. In some cases, fully automated systems are used to avoid or minimize human errors, but the interaction between planners, managers and operators should not be lacking for healthy decisions. New principles of work are to be designed. As an example, Human Supervisory Control Systems (HCS) used in power generation plants are digital and atomized for better management of the smart grid. We are still expecting the positive benefits of automated decisions, although the negative effect was shown a long time ago. Parasuraman and his colleagues studied central control cases and showed that fully automatic decisions have actually increased the operator's workload [PAR 00].

Talent hunting: demographic resistance limits the skilled operators of smart systems. Even if the user interfaces are very easy, a new way of training has to be given. It will be necessary to find ways for fast learning at work. Corporate social responsibility encourages hiring local talent. To conceive an adequate training program we need to know what we already have in term of competency and what is needed. A graphic such as knowledge trees is the visualization of available competence, or other decision systems are to be used for competence development management [ORD 15].

Optimized logistics powered by constraint programming and all other optimizing techniques help in sparing energy. Wind turbine maintenance is becoming an issue in the USA with almost 50,000 currently installed turbines. Automatic reliability checks and automated logistics with the spare parts can become a solution [JIN 12].

Significant challenges of ecological changes enforce every part of society to respond proactively including the manufacturing industry. There is a lot to contribute to manufacturing by providing efficiency in energy generations and material usages, and effective with the reduced energy use of the processes. Manufacturing system organization should consider the different levels of processing, which should manage individual devices and processing machines, and support equipment and the production facility.

Ecological application of Industry 4.0 can provide knowledge to achieve multiple objectives, eco-design, intelligent operation and management, and user-driven demand handling. Virtual engineering has become a new branch of engineering as a result of these operations. This new engineering enables the operation of smart machines, and creates and manages the experience base with semantic analysis and real-time visualization.

Emerging technologies include more enhancements in nanotechnology, biotechnology, robotics and humtech- the augmented human and combination of the above.

While all the above-mentioned technologies changed the way of working, it is also important to change the way the systems are designed and built. The expected last stage of Industry 4.0 is foreseen as virtual and real-time production with remote control. These changes cause the need for participative, predictive, preventive and personalized factories in the use of complex systems. The concept of 4P factory is developed to enable more effective use of tangible and intangible resources.

5.2.2. *4P factory*

The 4P factory is participative, predictive, preventive and personalized [AUF 10, HOO 12, 4PF 16].

Participative: the traditional top-down management is no longer effective in a hyper-connected world. Management methods are evolving from traditional theories to knowledge management, combining the top-down method supervised by bottom-up feedback and ideas. Emergent production management practices, artificial intelligence-based automation processes and product-driven lines require such approaches to manufacturing and industry.

Predictive: new products are eco-designed, simulated, 3D prototyped and produced with great reliance and reduced costs if we explore product line monitoring and past experience.

Preventive: prediction enables anticipation of potential problems and control of risks. It is knowledge- and experience-based. Simulators used before design help avoid wrong choices.

Personalized: clients and suppliers become stakeholders of the production process through customer feedback, small-loop logistics systems and even participation in the innovation process. Individual needs are served through high-volume production lines able to answer individualized orders.

The 4P factory requires switching models, paradigms, and changing minds. Global short-time logistic flows are required to merge with global knowledge flows. Aberkane [ABE 14] suggests the structuring of a knowledge-driven production according to three rules:

– knowledge gathering makes it more valuable, and is the condition for the predictive factory;

– knowledge transfer needs time and flow, so that the preventive factory can become a reality;

– knowledge exchange increases the resources of all exchange partners, and fosters the participative and personalized factory.

In fact, the whole product's lifecycle should follow the rules of holistic knowledge management applied to the extended business network of a given company supported by the right technology [AMI 97, MER 11]. The customer is involved in the company innovation process – he/she has also ideas and experience to share.

The main paradox remains – the quality of products supposed to be perfect is still strongly influenced by quick business and planned obsolescence.

These systems still require energy to work. There is no alternative solution proposed. In the case of a lack of energy or if the energy system is hacked by terrorists, the whole industry will be paralyzed.

The greatest challenge of industrial renewal is to consider all the impacts of innovation for sustainable success. These impacts follow system logic and influence each other. While eco-design focuses on the environmental impact, the other impacts are weakly considered. The objective of the Sustainability Chief Officer and the FCO (Financial Chief Officer) are not the same, while the overall company's objective is to become and remain a leader. The design department wants to achieve best performance compared with competitors. The final user role is just to buy. A more holistic and system view of industry connected to other activities should be considered.

5.3. From greening to smarting and wisening

The Rio Summit [RIO 92] initiated a global concern about planet preservation and put this topic at the governmental level. It proposed some conditions for reducing the impact of human activity on climate change without, however, setting the metrics for measuring the progress. Considering the multiple interests of participants, these objectives are very ambitious and difficult to achieve without changing the priorities and motivations both at the national and individual levels.

Corporate social responsibility defined the rules for companies that are constrained to respect the balance of three components: economic, environmental and social. Companies have to follow new norms: ISO 26000, REACH and others.

COP 21, held in Paris [COP 15], focused on CO_2 reduction and was directly related to the ways of producing and using energy.

Political incentives aim at promoting “clean energies” – solar, wind and other alternative energies are explored mainly separately, but some synergies are looked at. The combination of several sources of renewable energy is not evident because the companies are specialized and sell one. A typical example of absurdity is in Greece, where people always had a container on the roof to heat water in the Sun; now they are using containers and solar cells to do the same.

Companies exploring energy are searching for new sources, such as shale gas in the USA and China. The consequences of such exploration should also be evaluated before commercial exploration. Lack of such evaluation resulted in precautionary principles in France related to commercial exploration of this resource.

5.3.1. *From knowledge city to wise city*

The idea of knowledge city probably originated during 2003 at Entovation network meeting in Helsinki. En2polis is defined [ENT 03] as a physical and virtual space where world citizens are working together to share knowledge, experience and innovation. Such a city uses only the energy for powering the Website, but put together a lot of human energy, vital for innovating. The following year, the “Knowledge Cities, Knowledge Regions, Knowledge World” event and knowledge cafés were organized by El Tec, Monterrey, Mexico.

Since then, many initiatives around the world have flourished. In 2010, IBM launched a “Smarter planet” offer to reposition their products for eco-applications such as energy optimization or water management.

European programs on smart cities began in 2010 [SMA 10]. The initial idea is to use “digital technologies for better public services for citizens, better use of resources and less impact on the environment” (http://ec.europa.eu/digital-agenda/en/about-smart-cities). The main objective is to bring more ICT-based services to towns growing into Metropolis and Megapolis. This program focuses on zero energy buildings, optimized heating, cooling and energy as well as sustainable mobility. Other smart city programs are devoted to intelligent or green buildings, providing useful information to citizens on their well-being and environment.

Wise city is a recent initiative of the French Chamber of Commerce, in Hong Kong [WIS 16]. Among the initiators are large companies such as Dragages, the affiliate of Bouygues Construction, a global leader in the building, Alstom, a global leader in the world of power generation, power transmission and rail infrastructure, and BYME Engineering (HK) Ltd, a specialist contractor in all mechanical and electrical services and an expert in energy performance.

5.3.2. *Smart city and green city*

A city can be considered as "smart" when it wisely manages its tangible and intangible capitals, such as intellectual and social, traditional and digital communication, natural resources, quality of life, innovating continuously for the sustainable success of all participants. They learn from the past (accumulated knowledge) and from each other, solve day-to-day problems and improve their capacity. That implies a new kind of governance, involvement of citizens in public policy and the smart use of ICT.

For example, in Japan, companies such as IBM, Panasonic, Tokyo Gas and Accenture are focused on making solar energy a more crucial element in suburban planning. Is it enough to be a smart city? Giffinger *et al.* [GIF 07] states that a smart city can be ranked using six metrics, such as regional competitiveness, transport and digital economics, natural resources, human and social capital, quality of life, and participation of citizens in the governance of cities. The main objective is urban growth and development.

To face the economic crisis, cities and territories have to optimize their operations and reduce their costs. A smart city is an intensive user of ICT, intelligent technology, big data, connected objects and others. The smart city of Issy-les-Moulineaux also includes a smart energy grid [ISS 15].

Another current focus is on greening – more use of renewable energy, sorting garbage, recycling and eco-design. Greening is also concerned with green cities, green houses and green materials.

Green cities promote "clean" transportation, but we still need the energy to power them. Schneider's approach to a smart city is very interesting because it considers the whole system. Its architecture is shown in Figure 5.6.

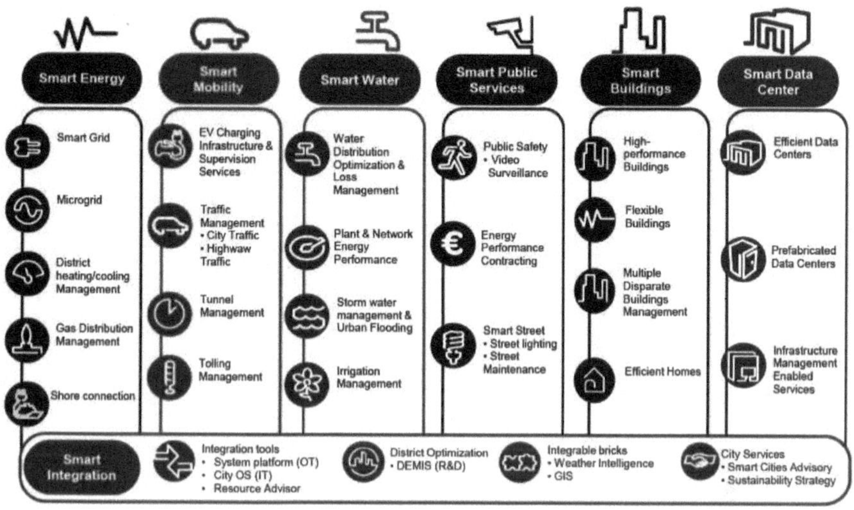

Figure 5.6. *Schneider Electric components of a smart city [LAR 15]*

According to Schneider Electronics, energy consumption in exponentially expanding cities will grow up to 75% by 2050. They use grids to manage the various facets of energy in the various contexts of a city.

The general principle of green cities covers architecture, pollution control and city gardening and farming. As an example, green and natural forms inspired the French architect Vincent Callebaut. His buildings are conceived to minimize energy required for heating and cooling. Plants are integrated into design and play a special role in maintaining the desired temperature.

Qatar Green Building Council aims at changing the modern architecture to a green one by the way buildings are conceived and used. Biotecture ambition is to develop environmentally sustainable and aesthetically pleasing cities with plants and vegetation that can operate as insulation on outer cladding to keep cool air in and heat out in summer. There is no question of cost, but of a different thinking.

Figure 5.7. *Lilypads by Vincent Callebaut*

Another example from Qatar is the campus at Education City and Mobility Innovations Center. The intention is to gather information about air quality. The pilot program involved multi-gas monitoring stations being placed at two locations on the campus. Information is sent to an intelligent sensing and service delivery platform that provides real-time air quality information. The e-survey aims to gather information from Qatari citizens and expatriates living in Qatar about how they use electricity, gas and water, in addition to their behaviors in using regular household appliances such as air conditioners, washing machines and televisions.

To protect and learn more about the dugong population of Qatar, a recent tri-party agreement was signed between ExxonMobil Research Qatar (EMRQ), a tenant of Qatar Science & Technology Park, Qatar University (QU), and Texas A&M University at Galveston (http://www.tamug.edu/).

The Petroleum Engineering program at Texas A&M University helps Qatar's ambitions in leading the way to safer, more environmental-friendly wells for the oil and gas industry. Solar energy is also in the spot. The roadmap outlines 10 conditions and related policy recommendations as a framework to discuss a range of inter-linked policies and actions to transform into a green economy. These conditions include social, environment and economic innovation; collaboration between all sectors; integrated governance; and balancing short- and long-term strategies and multilateralism. The roadmap also shares existing best practices and initiates new collaborative activities. The education of "green" attitude is on the way; the next step may be the education of "knowledge cultivators".

5.3.2.1. *Connected homes*

There are two approaches in connected homes: connected household appliances managed through internet and those who use connected utilities in a residential or industrial site.

The first one is obviously more popular since the residents have more responsibilities when the electrical appliances increase in number and the utility usage continues to go out of control. Safety and security seems to be the first reason of interest in connected homes, especially the fact that all the equipment is controlled when the residents are away. Socializing is presented as another reason with the IOT. Female residents have concerns about privacy. Even with the home hub, connectivity is not advised for those families who have limited electronic utilities at home.

The residential sites are connected mainly on all utilities. Since broadband has become the fourth essential utility after water, gas and power, equipment which was not afforded by the old properties are connected. The first results of Connect Ohio project were collected in 2013, which showed that already 72% of the state residents subscribed for broadband with remarkable energy saving in total.

New announcements for the connected homes have been made in most of the technology fairs this year. A new longer-range and low-power WI-FI standard is announced only for the connected homes. Samsung has announced the integration of a smart television and the fridge. Behind all these new products, there is a significant technology: ambient sensing.

Figure 5.8. *Example of connected homes in Dubai*

The cheaper technology gets, the more we see connected cars, connected kitchen devices, connected entertainments in the market. Anthropomorphic robots (remotely controlled humanoid robots) for home or business use are complementary to the connected home equipment. The wearable market is growing to more than one million units. All these new products use ambient sensing to reduce the future energy use. Connected products are designed to facilitate a new comfort definition that does not avoid communication and entertainment everywhere with a control of heat and power consumption [BAR 14]. Comfort is not necessarily healthcare or well-being, since those homes are permanently under the influences of an electromagnetic field.

A McKinsey survey shows that the primary buyers of connected devices are 25–44-year-old families with one to two children and the most preferred devices are the connected thermostat and multi-room music sets www.mckinsey.com/spContent/connected_homes/index.html.

Connected homes and products are important for energy efficiency and the health of residents, but they are also influential on the smart microgrids. Central control is no longer possible for power distributers of smart grids since the pieces are connected to each other as well as being connected to the service center.

How far will this architecture change the desert and other delicate ecosystems and biodiversity in the city? How does it influence an evolution of outdoor living without air conditioning?

Many other experiments are conducted around the world. UpoWa, a French start-up, provides for free small solar battery to people in Cameroon to help them light their house and charge their mobile phones (see Figure 5.9). In exchange, they collect data, information and knowledge about the other usages of the provided battery. This input is used to evolve their offer and to innovate.

Figure 5.9. *UpoWa solar battery*

Passive buildings or passive house is a construction concept aiming to optimize the energy use inside a building or a house. The libraries of models are proposed by [PAS 16].

5.3.2.2. *Circular energy*

Web 2.0 services generate a huge amount of multimedia data everyday that is stored in datacenters. Datacenters need powering and generate heat. Apple, Google, Cisco and Amazon all power their operations with energy recovered from datacenters that could supply our houses, shops and public facilities with heat and hot water. So far, this energy was simply evacuated into the air by means of different air conditioning systems. However, in recent years, recovery and reuse initiatives of these energy are in place to heat homes, offices, factories, swimming pools or even vegetable greenhouses.

On the Campus of Amsterdam, the Netherlands, the university has access to free hot water from the data center of Equinix that shares its campus. In Roubaix (north of France), the five OVH datacenters heat the company's offices and some neighboring firms. The data center of the British operator Global Switch also powers a tropical flowers greenhouse in Clichy, near Paris. In Uitikon, Switzerland, a public swimming pool is heated with an IBM data center. But the most impressive experience in Europe is conducted in the business cluster Val d'Europe, near Paris and Eurodisney. The heat produced by the Natixis bank datacenter is supplied to a district heating network. Hot water from the condensers of air conditioning systems is recovered by heat exchangers, and then distributed over the network using a central energy production. Ultimately, this system circulates water at 55°C in a network of 4 km of pipes, connected to the various buildings.

Aquaponic house: the principle of intelligent use of natural ecosystem properties is a circular architecture shown in Figure 5.10.

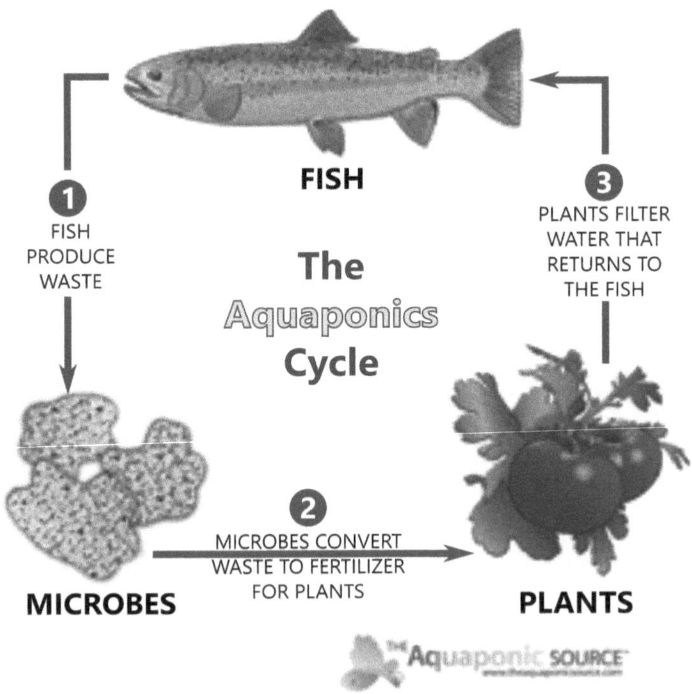

Figure 5.10. *Principles of the aquaponic cycle. For a color version of the figure, see www.iste.co.uk/kayakutlu/energy.zip*

Aquaponics is the marriage of aquaculture (raising fish) and hydroponics (the soil-less growing of plants) that grows fish and plants together in one integrated system. The fish waste provides an organic food source for the growing plants, and the plants provide a natural filter for the water the fish live in. The third participants are the microbes (nitrifying bacteria) and composting red worms that thrive in the growing media. They do the job of converting ammonia from the fish waste first into nitrites, then into nitrates and the solids into vermicompost that are food for the plants (http://www.theaquaponicsource.com)

Humanity will continue to hang on to the desire of having longer lives. Ecological balance can help improve the local and global cultures to support realizing the desires of sustainability. Recognizing the differences in different cultures will help find solutions for damages and education. We are in need of newly designed education to monitor the planet and the energy. Hence, we are in search of huge synergy that can be created through the collaboration of social scientists, life scientists and physical scientists.

A new fire is needed. A fire without flame that will help our welfare and well-being; however, tasks to complete this mission are not trivial.

5.3.2.3. *Urban farms*

Urban farms are not new, which were initiated by factories renting a small surface for workers to allow them to produce food and relax. Cultivating, processing and distributing food has become popular in big cities. The objective is mainly social – to provide local food for those who need it. Now it is being extended to the business of city farms. Some architects propose the ingenious integrated system of building farming, which is concerned with energy economy. The Pasona HQ building in Tokyo is an example of such architecture, as shown in Figure 5.11.

This building of vertical farming, designed by Kono Design, has many objectives: to provide a nice façade and interior, provide fresh food for employees and reduce total energy use in the building. Green facades also protect and isolate.

Figure 5.11. *Pasona interior and exterior views*

5.4. Managing the uncertainty of dynamic demand

Global warming, fossil energy scarcity and the political independence of regions or countries are the reasons which lead to increasing the use of renewable energy sources. Thus, the list of uncertainties faced in energy operations is opened to new challenges. Demand and price fluctuations are widely accepted uncertainties as in any industry. Political decisions to support certain resources or provide subventions for certain energy sources are risks to handle in economic analysis. Stochastic and/or fuzzy models are designed to operate in uncertain environments. But now, the variability of wind or solar energy availability causing changes in conventional generator outages is introduced [FIT 15]. Moreover, energy generation using renewable energy resources has the concern of output fluctuation. Dynamic and hybrid models help to decrease the mistakes in predictions and fight against the indeterministic schedules.

A recent study by Audun Josang combines probabilistic analysis, Bayesian belief network and fuzzy approach in subjective logic [JOS 14].

This new approach enables a trusted interaction between Internet and mobile applications for degrees of uncertainty where belief of ownership can be expressed. Uncertain modeling and analysis situations and incomplete knowledge can benefit subjective logic.

Unique focus on renewable energy reduces pollution, but does not create a huge novelty. As Glasnovic *et al.* [GLA 15] mentioned using renewable energy resources, the ability to store generated power and install smart grids will not change industrial life a lot. The real revolution will be recycling the power used by all the machines and technical utilities not only in energy resources, but also in water and air sources.

The algorithms and methods supporting the dynamic energy environment can be deterministic to stochastic, periodical to sequential, static to dynamic, discrete to continuous and mono-agent to multi-agents, but certainly complex systems. A complex system is typically adaptive or evolutionary and influenced by social and political, as well as physical, processes.

The synergy between the complex systems and the energy systems is well established by using the hybridization of engineering, biological, social and physical sciences. The systems approach allows the management of the whole through the management of the components. System dynamics allows defining the components and the influences of the positive or negative feedbacks at any nested level. Energy systems are to be understood through alternative ways such as:

– *Case-based learning*: Germany has made a leap through the full solar use in certain districts, which would be a case-study for developing countries like Turkey.

– *Cultural education*: after the Paris Climate Change Conference on November 2015, everybody recognized that a country is not only changing the local ecology but disrupting the natural equilibrium across the globe. Hence, a new style of education is to be designed to improve the energy knowledge.

– *Collaborative knowledge use*: technologies and new energy resources are innovated to accelerate the developments in the energy field. Individual companies, governments and research institutes are to collaborate with the aim of creating solutions for complex forecasts and plans.

Ulrich Beck has written admirably on the role of knowledge in organizing present activity [MAS 06]. He notes that the relevance of a projected cause, a projected variable and a projected outcome is directly proportional to their perceived threat. Thus, the future with notions of progress is synonymous with risks. Efforts to manage risk have become an open-ended, future-oriented project, the goal of which is to anticipate all the possible forms of uncertainties.

5.4.1. *Satellite images*

Satellite images are mainly used to control larger areas like forests, lakes, rivers and mainly urban areas. They serve in energy-efficiency improvement by following traffic and changing routes or in taking precautions against urban pollution. Observing the lakes and forests provides information about humidity and meteorological, moving predictions closer to reality.

In urban parks, beaches, rivers and the like are stated as meteorological, human interventions and the social activities. Organized events or visits are to be observed in correlation with cultural reactions. Images are also used to measure the performance of PV panels or wind turbines based on the humidity, temperature and wind speed [BAL 15a, BAL 15b]. Predicting wave movements is important for both storms and wave energy.

Such images are used for smartphone applications for smart implementations in a city or even in a district, as anything searched can be spotted. Improvement of urban environmental conditions is made possible. Furthermore, finding the most convenient places for solar panels can increase energy performances. Rainwater retention can be closely followed. A study in Thessaloniki shows remarkable variation in the city even if it is not a metropolis. Roof availability for solar panels is detected in addition to defining the water retention in central old apartments. Rain water retention is increased by using green roofs.

Smart cities are following the heat islands to predict cooling demand or storm arrivals to predict heating demand. Unless pictures are used, data is collected from different geographical parts of the city. In Istanbul, it rains heavily in the northwest of the European side, whereas it is cloudy at the European sea side and hot and sunny on the Anatolian side. The whole 360° picture of the urban area can be taken in a few minutes using this method. Pictures are advantageous for predicting the humidity, temperature and tides whenever there is water in the city, and the dust flow in deserts. These data are used in both energy management and agricultural improvements.

5.4.2. *Zero energy*

Ultimate need of "zero energy consumption" is already defined by US Department of energy as follows [DOE 15]:

– *Zero energy building*: an energy-efficient building where, on a source energy basis, the actual annual delivered energy is less than or equal to the on-site renewable exported energy.

– *Zero energy campus*: an energy-efficient campus where, on a source energy basis, the actual annual delivered energy is less than or equal to the on-site renewable exported energy.

– *Zero energy portfolio*: an energy-efficient portfolio where, on a source energy basis, the actual annual delivered energy is less than or equal to the on-site renewable exported energy.

– *Zero energy community*: an energy-efficient community where, on a source energy basis, the actual annual delivered energy is less than or equal to the on-site renewable exported energy.

As it is observed, the definition is focused on efficiency not necessarily on any creativity.

One promising solution that has gained attention recently is incorporating free and natural resources from nature such as lighting and natural air. In terms of ventilation, the ultimate aim would be to achieve a mixed mode of natural ventilation and air conditioning at different times of the day, year or season in order to provide a satisfactory comfort level for the occupants and a healthy indoor environment. Natural ventilation is already used by architects [HER 10].

5.4.3. *Energy storage*

The old dream of harnessing the storm and storing this energy is not yet a reality, but the ability to store electricity generated by using renewable resources has reduced uncertainties.

The current storage techniques are enabled by power electronics chemistry and intelligent integrated engineering. With electric cars, storage has become easier [BLA 05], even making the cars themselves the storage media.

Sony Computer Sciences Laboratory Tokyo is working, among others, on innovative Open Energy Systems (OESs) that redefine the energy distribution from the top-down grid to interconnected systems. The three pillars of OES are as follows:

– *Sustainable*: managing renewable energy sources that are geographically distributed and of which the output reserved is granted to batteries.

– *Dependable*: exploiting locality of generation and consumption and increasing self-sufficiency by interconnecting a system in a bottom-up and autonomous way;

– *Affordable:* starting with a small yet expandable system in a bottom-up, autonomous and interconnected way [SON 15a].

They conduct several experiments on such systems around the world [SON 15b].

A white paper from the Rocky Mountain Institute defines energy storage batteries to serve utility companies, energy distribution companies and the consumers. Transmission companies use it for resource adequacy, and they follow transmission deferral abilities. The service companies try to regulate the power frequency spinning and the voltage support, which are the most important features. The consumers use batteries as a backup unit, hence photovoltaic abilities are the focus of interest [ROC 16]. Energy use shifting will be applied for the demand-side management. Several studies proved that in connected districts, cooperative energy use will be more beneficial for all residents.

5.4.4. *Smart grids*

In the beginning of the 20th Century, district heating started to build residences in collective construction in Denmark. With the improvements in electricity demand, they have become local cogeneration centers using natural gas or oil. Cooling systems are introduced once the hybrid renewable energy sources are used.

A Swiss Project "Vision of Future Energy Networks", launched in 2007 by ETH Zurich, found two key reasonable approaches: transformation, conversion, and storage of various forms of energy in centralized units called energy hubs, and combined transportation of different energy carriers over longer distances in single transmission devices called energy interconnectors. The achievements were expected to connect the existing systems to the optimal structures [GEI 07]. These predictions were handled in smaller demand centers.

The energy demand for a certain district or center was responded to by a few central generators. The use of renewable energy sources caused a paradigm shift as well as high technology use in grid controls. Nowadays, sizes grid are getting smaller (can even be for a single residence) because of fast improvements in renewable energy plants. In addition to pollution decreases and supply diversification, new job developments in the installation and manufacturing of renewable generation units and grid operations facilitate the social acceptance of the grids.

Microgrids are key elements to integrate renewable and distributed energy resources as well as distributed energy storage systems. A central operation and smart measures of microgrids are easily acceptable for the residential sites, but not widely spread among industrial or commercial use due to the variability of the resources and uncertainties. A key feature of microgrids with distributed energy sources is that the sources are dispersed over a wide area. These sources are interconnected to each other and to loads by a distribution network. Furthermore, the distributed microgrid may be connected to the main power grid at some point as well.

Dispatching of the central energy generation is to be planned by considering the co-operative benefits, and the advantages over the individual targeting are to be shown. Recent technologies in energy storage with more powerful batteries such as fly-wheel, or metallic oil, make smart grids more powerful with certain extra units of energy kept to recover the uncertainties. Today, even plug-in of electrical vehicles to the grid is possible. The central services are developed to take over the responsibility of not only the technical operation of generation and dispatching, but also the administrative work load of measuring, scaling and billing.

Finding the solution for the economic and environmental solutions in small and medium grids is handled as Demand Site Management (DSM). Thanks to the DSM software, the following can be adopted to the grid:

Power can be generated or bought from outside the grid through liberal retailers or bigger grids.

Both smart measuring technologies and ambient sensing can be used to combine consumption data and consuming behavior. Knowledge-based analyses are performed to change the production plans to avoid fluctuations in the energy balance. SCADA and the energy management system work together with the ambient simulators and the weather forecasts.

Using energy storage units and power charging units gives the opportunity to sell extra units to companies outside the grid. Different tariffs of the day-ahead calculations and spot responses are to be managed by the service center.

Grid management becomes a complex optimization center to guarantee the equality of supplies and demands.

5.4.5. *Center of service*

Smart district has a central unit, which can be called the center of maintenance, center of service or center of distribution. The major responsibility of that center is to track all the smart systems while responding to the demand in continuous availability.

Considering the concepts of Industry 4.0, there are eight main duties of center of service, as shown in Figure 5.12.

Demand response and secure continuity are the basic features. All the required energy by the district has to be supplied in a continuous security that can only be possible with optimized distribution:

– *Islanding*: this is the ability to generate power by combining the distributed resources even when utility is absent. Rooftops, fuel cells and electric vehicles should guarantee that security centers or health centers are up and running.

– *Peak shaving*: by using thermal and/or battery storage options, peak usages can be reduced. The lower the peak periods, the cheaper the cost of utility. This requires an optimized scheduling of all generation and storage options.

– *Smart measures*: fast collection of data is possible with the sensors, but only the analysis of the data to support fast decisions makes it smart.

– *Smart sensors*: smart agents are used in buildings and generators to enable the knowledge accumulation in pictures or video, in voice and or in heat detecting.

– *Self-healing*: system defects and disturbances have to be detected and responded to fast. With the focus on prevention, the consumer impact would be reduced.

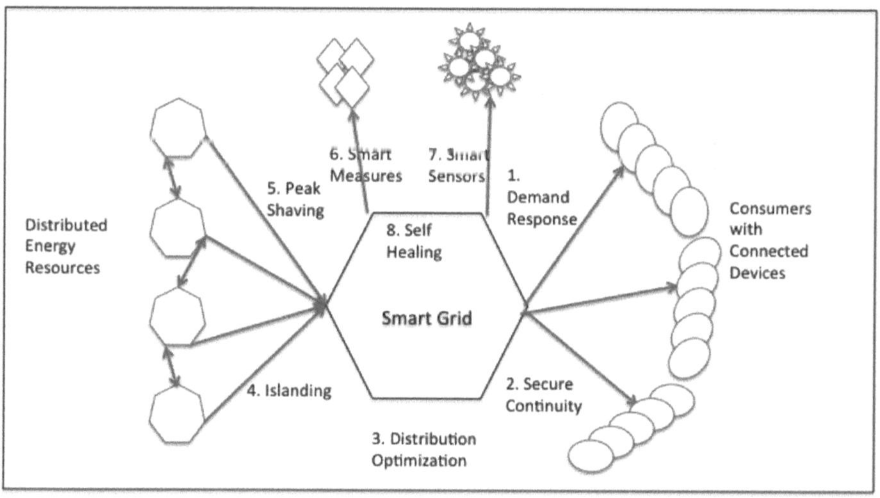

Figure 5.12. *Center of maintenance with eight duties*

The US Department of Energy lists five fundamental technologies that will drive the smart grid [DOE 16]:

– integrated communications, connecting components to open architecture for real-time information and control, to have interactions of all components;

– sensing and measurement technologies, to support faster and more accurate response such as remote monitoring, time-of-use pricing and demand-side management;

– advanced components, to apply the latest research in superconductivity, storage, power electronics and diagnostics;

– advanced control methods, to monitor essential components, enabling rapid diagnosis and precise solutions appropriate to any event;

– improved interfaces and decision support, to amplify human decision-making, transforming grid operators and managers quite literally into visionaries when it comes to seeing into their systems.

5.5. From collective to individual – some trends and experiments

After collective exploring of energy, the trend is to be autonomous. Small planes for one or two persons are developed by Airbus research, and individual cars are able to run 1,000 km with only 1 liter of a biofuel that are designed by the students of Ecole Polytechnique Federale de Lausanne and Ecole Centrale Lyon. Individual houses become autonomous and combine all available energies. Circular energy systems are prototyped; among them, the Datacenter of Google in the north of Finland exists with the aim of heating the campus buildings with hot water from DC cooling.

5.5.1. *New transportation*

A lot of effort is made in smart transportation, but little investment is made in new means of transportation known from science fiction books.

The students of EPFL and ECLyon organize each year a competition for cars able to ride long distance using only a liter of biofuel/1,000 km.

An average realistic lifespan of a mobile phone is 1.5–2 years and the batteries used in a mobile phone have a lifecycle of 1,500–2,000; with a single charge a day, the battery still has more than 50% of life when sent to recycling [DIO 15]. Those batteries are currently used in the lightning of small units. Batteries of electric cars are more powerful; technologies are on the way to use them for residential heating or cooling after they lose power for the cars.

The electric vehicle can be a quantum leap in transportation from local contamination reduction to new industrial opportunities [MAR 13]. The revolution in circulation in vehicles will not only be the usage of batteries but also using self-energy creation through the engine as soon as it starts running. The waste gas usage was made as a novelty [HU 15, MIT 11].

Electric cars are so well accepted in some societies that Nissan launched a project for electric cars to store solar/wind energy and distribute power around the connected districts [MON 16]. Cars could be parked next to a home at night and pump energy into the house and when driven to work the next day, it could be used to power offices. This project is run with architects to create smarter cities. A world bank statistics shows that by 2050, 54% of the global population will live in cities and 70% of those will prefer urban areas.

5.5.2. *Solar impulse*

Solar impulse is exclusively sun energy-powered plane (see Figure 5.13). It is able to fly around the world. Currently, it can transport only one person, but we hope it will be more powerful in the near future.

Figure 5.13. *Plane flying with solar energy [SOL 15]*

There are also boats powered by solar energy. One of them has been floating in the oceans since April 2015 to support a Suisse Foundation Race for Water [PLA 15]. The foundation is against the plastic pollution in the oceans.

Figure 5.14. *Boat floating with solar energy [PLA 15]*

5.6. New trends

Recent studies have shown that clean energies are replacing fossil sources in many countries (except India where coal is on the rise) in addition to discovering new resources. Commercial and industrial users are taking the lead in realizing the future objectives (e.g. Google purchased 2GW of clean energy). Moreover, after COP21, governments will increase the funds for research and development in the energy field. New finance resources for the future of energy are also allocated. The Breakthrough Energy Coalition (BEC) has been established among private investors to create funds of multi-billion dollars to bring new energy technologies to the market [FEL 16].

Smarter Planet is a corporate initiative of IBM connecting their existing offer and the planet concern. The goal is to compose/design "smart systems" able to manage energy, water, transportation in smart cities and generate revenue from open data. The initiative aims at connecting businesses, government and citizens around the common concern – sustainable development.

Examples of smarter systems include smart grids, water management systems, solutions to traffic congestion problems, greener buildings, flood prevention and many others. These systems are supposed to be difficult to manage because of their size and complexity. With new ways of monitoring, connecting and analysis, IBM offers new ways to manage them. Their strategy is to provide or enable some of the existing technologies (database, analytics, rules, solver) and process management capabilities to influence policy decisions in order to "make the planet smarter" and make smart business.

In the field of energy, their TRIRIGA software combines real-time monitoring with event management and analytics to help manage energy in buildings and improve maintenance and reliability [IBM 16].

5.6.1. *Ecosystem development*

Investors and entrepreneurs in energy technology are encouraged to develop an ecosystem from day one of their projects. The ecosystem is to cover early adopter customers, partners from the financial industries for funds and from the insurance industries for risk reduction, partners in product development and supply chain, and partners in independent engineering and certification. The investments have to create trust that they are investing for the benefits of humanity and the nature.

Mirova, Asset Management achieved The "Energy and Ecological Transition for Climate" label for the actual impact on climate change, water, natural resources and biodiversity with three projects: Eurofideme 3 is a renewable energy infrastructure fund for projects that produce energy from renewable resources in Europe (over the past 10 years financing 730 MW of new generating capacity helped avoid 1.4 million tons of CO_2 emissions)

Green Bond invests in securities supporting energy and ecological transition, and allows issuers to diversify their investor base and investors to actively participate in financing energy transition. Europe Equity Fund is a thematic equity fund. It invests in shares of European companies that are positively exposed to innovative environmental issues covering energy efficiency, food production cycle and water management [MIR 16].

Hydraulic fracturing (Hydrofracking) is a well-known technique of extracting the natural gas, first occurring in 1947. With this method, large amounts of water, combined with smaller amounts of chemicals and sand, are pumped under high pressure into a drilled gas well. The purpose of hydraulic fracturing is to form tiny fractures in the rock by using water to force the rock to open along tiny existing fractures. When the pressure is released and the water removed, the sand remains behind [WIL 16].

This method has brought the US energy independence high hopes, but the mitigation is prevented because of the harms listed below [HOF 12]:

– contamination of groundwater;

– methane pollution and its impact on climate change;

– air pollution impacts;

– exposure to toxic chemicals;

– blowouts due to gas explosion;

– waste disposal;

– large volume water use in water-deficient regions;

– fracking-induced earthquakes;

– workplace safety;

– infrastructure degradation.

5.6.2. *Bankable energy storage*

Power or heat storage is the remedy for renewable energy investments. However, bankability of energy storage is not yet ready due to the risks and low profits in the field. Bankability of a project requires the analysis and approval of legal, technical and economic issues in general, but includes ecological issues as well for the energy storage projects. Hence, new technologies are on the way to bring power storage the approval of integrated analysis of financial fund managers, regulators, ministries and project developers. Researchers are working on higher volume power storages and the possibility of keeping the hot water for longer than a couple of hours.

Compressed air energy storage (CAES) is a plant where ambient air is compressed and stored under pressure in an underground cavern. When electricity is required, the pressurized air is heated and expanded in an expansion turbine driving a generator for power production. The air heats strongly; that is why multi-stage air compressors use inter- and after-coolers. The compression stage uses up to about two-thirds of the gas turbine capacity. With current technologies heat loss is recaptured and is used at the compression stage, hence the efficiency is improved up to 70%. Very large storages are required with this method because of the low storage density [ENE 16].

5.6.2.1. *Hydrogen and fuel cell as a power storage*

Hydrogen is high in energy, yet an engine that burns pure hydrogen produces almost no pollution, and hydrogen can be separated from hydrocarbons like natural gas through the application of heat. An electrical current can also be used. Water electrolysis technology is a flexible solution to store renewable energy on a large, long-term scale. Using excess renewable electricity, the proton exchange membrane (PEM) electrolyzer splits water into its constituent parts, hydrogen and oxygen. [SCH 14].

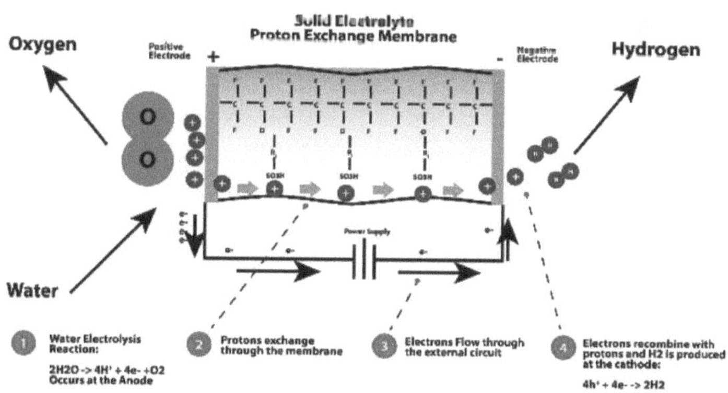

Figure 5.15. *Electrolysis to obtain hydrogen from water [SCH 14]*

Under certain conditions, hydrogen can even be retrieved from some algae and bacteria. Renewable Energy World states that (http://www.renewableenergyworld.com/index.html):

> "NASA has used liquid hydrogen since the 1970s to propel the space shuttle and other rockets into orbit. Hydrogen fuel cells power the shuttle's electrical systems, producing a clean byproduct – pure water, which the crew drinks. A fuel cell combines hydrogen and oxygen to produce electricity, heat, and water"

The fuel cell produces electricity as long as hydrogen is supplied, never losing its charge. Fuel cells are promising for the future as a source of heat and electricity for buildings, and as an electrical power source for electric vehicles. Fuel cells operate best on pure hydrogen.

In the future, hydrogen could also join electricity as an important energy carrier that moves and delivers energy in a usable form to consumers. This is the feature that makes hydrogen more interesting than the synthetic natural gas (SNG) that is produced using renewable energies, although it is found to be more efficient in storing power [CHA 09]. Hydrogen can be stored until it is needed and can be transported to locations where it is needed [REW 16].

5.6.2.2. *Battery for utility*

Current batteries use metallic oil like lithium and can store small amounts because of the very high cost. In fact, 70% of the world's lithium supply comes from limited sources of Chile, Bolivia and Argentina, commonly known as the "Lithium Triangle". Donald Sadoway and his team at Massachusetts Institute of Technology (MIT) have innovated a rechargeable battery of molten metals, which offers a low cost and long-lasting storage [STA 16]. Sadoway reversed the aluminum smelting process to use electricity produced, which is conducted inside electrochemical cells. The battery has top and bottom electrodes with electrolyte between them. Unlike classical batteries, the top layer, the negative electrode, is a low-density liquid metal; the bottom layer, the positive electrode, is a high-density liquid metal and in between, the middle layer molten salt transfers charged particles. Aluminum smelting and salt usage create a cheap operation with high-volume storage potential. The technique is not yet ready for the market.

5.6.2.3. *Plants for energy storage*

Phase changing materials are used for thermal energy storage [PHA 16]. Organic materials change phases at different heating degrees because of the combination of carbon and hydrogen. Plants are also organic material where hydrogen could be saved and the juice of the plant has a certain thermal feature. Could we use plants for energy storage by following the phase changes?

5.6.3. *New transportation*

After presenting the solar vehicles in the previous pages, there are more to come out of research laboratories.

5.6.3.1. *Hydrogen vehicles*

TESLA, the electric car producer, is expecting to sell 500,000 cars by 2020. Electric cars are not only energy efficient, but also clean energy users without exhaust pipes. Currently, the most efficient battery used for electric vehicles is lithium, which makes the car expensive. Hydrogen storage for vehicles was a project run between 2009 and 2015 in National Renewable Energy Laboratory, USA and is very close to market [THO 14]. The vehicle features are adopted to request hydrogen to the storage system and use fuel cells for the power flow. Hence, these cars use the electrolyzer to separate hydrogen from water. Thermal management is the key issue that is to be solved to enable secure use of hydrogen storage and power busses. Simulation of test models can be accessed at [IISE 16].

Sumitomo corp (www.sumitomocorp-global.com/) and US Hybrid corp (http://www.ushybrid.com/) collaborate on Hydrogen Fuel Cell Powered Public Transit [ENV 16]. The first model is a combined lithium battery (76–125 kW battery) and hydrogen storage (40 kg) that uses the fuel cell system (120 kW) [SOL 16].

The National Aeronautics and Space Administration (NASA) announced that electric planes will be in the sky in 2018. Electrical airplanes are already used. Boeing Fuel demonstrator and solar powered planes are showing up in the skies every other day. The NASA plane not only uses the fuel cells, but also has a glass cockpit, which has been recently developed [MUR 15].

Pilotless drones to transport humans were shown in a Consumer Electronics & Consumer Technology (CES) show in 2016 in Las Vegas,

USA. Ehang, the producer of that drone, state that they have taken all the security cautions like multiple power backups, a program that will land the drone at the first sign of trouble, and a detailed command center for fliers who need help [GOL 16]. Battery-powered drones use clean energy, but there are ecological concerns if it is not as big as a plane.

Hyperloop train, the supersonic transport system proposed by Elon Musk, has taken the first public tests in Nevada sponsored by many companies including French SNCF. This train works by propelling pods at high speeds through a tube, which in theory would be able to make the journey from San Francisco to Los Angeles in just 30 min. It is expected to create an alternative for using cars in geographically huge cities like Istanbul [WOO 16].

Human creativity has no limits. We expect many new "energy light" solutions in the near future.

Conclusion

Energy and the Future

Today, energy is managed by governments and businesses and is considered as a business and public service.

New forces ranging from persistently depressed crude oil prices to growing climate change policies are making most governments alter their forecasting approach. More frequent analysis and reports are preferred; they are produced following the discovery that long-term predictions are less useful for both energy suppliers and consumers. All these analysis and reports are based on "data thinking" and linear approximations.

Such predictions are not compatible with the current dynamic world. Radical changes may also come from some users aware of the intelligent use of energy.

New policies have emerged and been run both at the country and international levels to share the newly discovered resources. The International Energy Agency (IEA) has redefined its policy to serve worldwide instead of just working for the benefits of IEA members.

The IEA sees the future of energy as more uncertain than ever. The discussions focus on three foresights: (1) it is observed that with wider use of renewable energies and energy saving, supply has become greater than the demand; (2) future oil and gas levels are subject to uncertainty because of the prices and less investments on pipelines; (3) changing energy sources in China and India may cause unexpected growth in demand for all players in the energy supply chain.

Even governors admit that policies should be based on technologies which have never been developed but have great importance in the future with the focus on raising energy awareness. If citizens cannot interact with policies, then politicians will continue to control the energy future.

Environmental constraints are an extraordinary motivation to innovate in all fields related to energy. Progress is made with the combination of traditional and renewable energies. The next step is to better understand how to use other natural energy sources.

In this respect, we often forget the energy of connected human brains collaborating in synergy with connected computers empowered with "intelligent" techniques. They are able to propose balanced solutions. In fact, we expect the new technology suggestions of wise and frugal use of energy respecting the balance of the ecosystem.

Today, energy ecosystems lack balance because decisions are taken only between governments and energy producers.

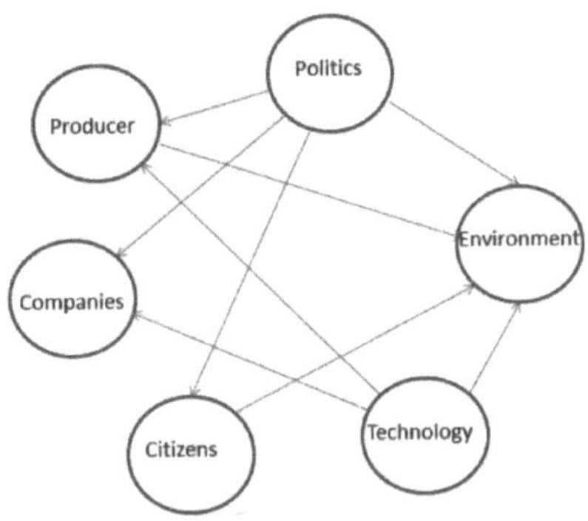

Energy users such as companies, organizations and individuals have to pay taxes and are not asked for feedback and ideas. Despite governmental decisions on CO_2 reduction, all have a heavy impact on the environment. This impact is reflected in natural disasters that are more numerous and more powerful.

To preserve this balance, new connections are needed:

– sharing the innovative individual practices;

– providing feedback to governments;

– eco-designed technology.

We hope the "wisening" of:

– energy users;

– energy providers with less focus on business and more on the environment and well-being;

– researchers with more focus on understanding solutions provided by nature and evaluating the impact of their findings;

– technology producers evolving into eco-designers;

– governments innovating with citizens.

Bibliography

[A4C 16] A4CP, available at: http://www.a4cp.org/, accessed in June 2016.

[ABE 14] ABERKANE I., "Black gold, green gold, blue gold: why is knowledge economy our next renaissance?", at *csdc-wiki.unistra.fr*, 5 June 2014.

[AGU 12] AGUSTONI A., MARETTI M., "Energy and social change: an introduction", *International Review of Sociology*, vol. 22, no. 3, pp. 391–404, 2012.

[ALP 11] ALPAYDIN E., "Machine learning", *Computational Statistics*, vol. 3, no. 3, pp. 195–203, 2011.

[ALT 10] ALTAY A., KAYAKUTLU G., TOPCU İ.Y., "Win-win match using a genetic algorithm", *Applied Mathematical Modelling*, vol. 34, pp. 2749–2762, 2010.

[AMI 97] AMIDON D., *The Innovation Strategy for the Knowledge Economy*, Butterworth Heinemann, Boston, 1997.

[ARI 14] ARIGLIANO A. *et al.*, "Producing, storing, using and selling renewable energy: the best mix for the small medium industry", *Computers in Industry*, vol. 65, no. 3, pp. 408–418, 2014.

[ASL 13] ASLAN A., "Energy consumption and GDP: the strong relationship in OECD countries", *Energy Sources, Part B: Economics, Planning, and Policy*, vol. 8, no. 4, pp. 339–345, 2013.

[AUF 10] AUFFRAY C., CHARRON D., HOOD L., "Predicitive, preventive, personalized and participatory medecine – back to the future", *Genome Medecine*, vol. 2, no, 8, pp. 57, 2010.

[AUR 05] AURIOL E., MANAGO M., ALTHOFF K.-D. *et al.*, "Integrating induction and case-based reasoning: methodological approach and first evaluations, advances in case-based reasoning", *Lecture Notes in Computer Science*, vol. 984, pp. 18–32, 2005.

[AWE 93] AWERBUCH, S., "The surprising role of risk in utility integrated resource planning", *The Electricity Journal*, vol. 6, no. 3, pp. 20–33, 1993.

[AWE 06] AWERBUCH S., "Portfolio-based electricity generation planning: policy implications for renewables and energy security", *Mitigation and Adaptation Strategies for Global Change*, vol. 11, no. 3, pp. 693–710, 2006.

[AZA 15] AZADEH A., SOHRABI P., SABERI M., "A unique meta-heuristic algorithm for optimization of electricity consumption in energy-intensive industries with stochastic inputs", *International Journal of Advanced Manufacturing Technologies*, vol. 78, no. 9, pp. 1691–1703, 2015.

[BÄC 97] BÄCK T., FOGEL D., MICHALEWICZ Z., *Handbook of Evolutionary Computation*, Oxford University Press, 1997.

[BAC 13] BACCINI L., LENZI V., THURNER P.W., "Global energy governance: trade, infrastructure, and the diffusion of international organizations", *International Interactions*, vol. 39, pp. 192–216, 2013.

[BAL 15a] BALDI S., KARAGEVREKIS A., MICHAILIDIS I.T. *et al.*, "Joint energy demand and thermal comfort optimization in photovoltaic-equipped interconnected microgrids", *Energy Conversion and Management*, vol. 101, pp. 352–363, 2015.

[BAL 15b] BALE C.S.E., VARGA L., FOXON T.J., "Energy and complexity: new ways forward", *Applied Energy*, vol. 138, pp. 150–159, 2015.

[BAN 98] BANZHAF W., NORDIN P., KELLER R. *et al.*, *Genetic Programming – An Introduction*, Morgan Kaufmann, San Francisco, 1998.

[BAÑ 11] BAÑOS R. *et al.*, "Optimization methods applied to renewable and sustainable energy: a review", *Renewable and Sustainable Energy Reviews*, vol. 15, no. 4, pp. 1753–1766, 2011.

[BAR 99] BARANWAL A.K. *et al.*, "Aldosterone-producing adrenocortical adenoma in childhood: a case report", *Journal of Pediatric Surgery*, vol. 34, no. 12, pp. 1878–1880, 1999.

[BAR 14] BARBATO A., CAPONE A., "Optimization models and methods for demand-side management of residential users: A survey" *Energies*, vol. 7, no. 9, pp. 5787–5824, 2014.

[BAR 04] BARRETO L., KYPREOS S., "Endogenizing R&D and market experience in the "bottom-up" energy-systems ERIS model", *Technovation*, vol. 24, no. 8, pp. 615–629, 2004.

[BAR 11] BARTHALMIE R.J., PRYOR S.C., "Meteorology and wind resource assessment for wind farm development", *Wind Energy Systems*, pp. 3–27, e28, 2011.

[BAS 82] BASILE P.S. *et al.*, "The IIASA set of energy models: its design and application", in *Energy Modelling Studies and Conservation*, Pergamon Press, Oxford, pp. 87–126, 1982.

[BAZ 93] BAZARAA M.S., SHERALI H.D., SHETTY C.M., *Nonlinear Programming: Theory and Algorithms*, 2nd ed., Wiley Interscience, New York, 1993.

[BEJ 96] BEJAN A., "Method of entropy generation minimization, or modeling and optimization based on combined heat transfer and thermodynamics", *Revue Générale de Thermique*, vol. 35, nos 418–419, pp. 637–646, 1996.

[BER 90] BERNDT E.R., "Energy use, technical progress and productivity growth : a survey of economic issues", *Journal of Productivity Analysis*, vol. 2, pp. 67–83, 1990.

[BER 99] BERTSEKAS D.P., *Nonlinear Programming*, 2nd ed., Athena Scientific, 1999.

[BER 03] BERGMANN R., ALTHOFF K.D., BREEN S. *et al.*, "Developing industrial case-based reasoning applications. The INRECA methodology", *Lecture Notes in Artificial Intelligence*, 2nd ed., vol. 1612, Buchreihe, Springer-Verlag, Berlin, 2003.

[BHA 12] BHATTACHARYA A., KOJIMA S., "Power sector investment risk and renewable energy: a Japanese case study using portfolio risk optimization method", *Energy Policy*, vol. 40, no. 1, pp. 69–80, 2012.

[BLA 05] BLAABJERG F. *et al.*, "The future of electronic power processing and conversion", *IEEE Transactions on Industry Applications*, vol. 41, no. 1, pp. 3–8, 2005.

[BLE 07] BLESL M. *et al.*, "Role of energy efficiency standards in reducing CO_2 emissions in Germany: an assessment with TIMES", *Energy Policy*, vol. 35, no. 2, pp. 772–785, 2007.

[BON 86] BONNET A., HATON J.-P., TRUONG-NGOC J.-M., *Systèmes Experts – vers la maîtrise technique*, InterEditions, 1986.

[BOR 16] BORNSCHLEGL M., BREGULLA M., FRANKE J., "Methods-energy measurement – an approach for sustainable energy planning of manufacturing technologies", *Journal of Cleaner Production*, vol. 135, pp. 644–656, 2016.

[BOU 14] BOUDON R., REBUT P., MONNERIE J. *et al.*, "Utilisation des drônes pour la surveillance des ouvrages de production d'EDF dans le domaine de génie civil", *Revue* XYZ, no. 141, pp. 55–60, available at: aftopo.org/download.php?matricule=414111, 2014.

[BRO 13] BROOKINGS REPORT, A trilateral dialogue on the United States, Africa and China, available at: https://www.brookings.edu/wp-content/uploads/2016/07/All-China-Oil-Papers-2.pdf, 2013.

[BRU 02] BRUNO K., KARLINER J., "The UN's global compact, corporate accountability and the Johannesburg Earth Summit", *Development*, vol. 45, no. 3, pp. 33–38, 2002.

[BUR 01] BURBAGE-ATTER M.G., "Energy conservation", in MOBLEY K. (ed), *Plant Engineer's Handbook*, Chapter 30, pp. 503–516, Butterworth-Heinemann 2001.

[BUR 11] BURGER O.D., HAMILTON M.J., "*Industrial energy use and the human life history*", *Scientific Reports*, vol. 1, no. 56, 2011.

[BUZ 03] BUZAN T., *The Illustrated Mind Map Book*, BBC, 2003.

[CAH 12] CAHILL C.J., GALLACHÓIR B.P.O., "Quantifying the savings of an industry energy efficiency programme", *Energy Efficiency*, vol. 5, no. 2, pp. 211–224, 2012.

[CAM 12] CAMPIGLIO E., BERNARDO G., "Modelling the great transition", *International Conference on Ecological Economics, Rio de Janairo*, available at: http://www.slideshare.net/emacampiglio/nef-macro-model-isee-rio-2012, 2012.

[CAM 14] CAMPBEL J., "Center pivot system for data collection for agriculture, has monitor that receives energy signal from transport device and signal powers wireless signal from transceiver of monitors to transceiver of transport device", US2014024313-A1, 2014.

[CAN 14] CANO E.L. *et al.*, "A strategic optimization model for energy systems planning", *Energy and Buildings*, vol. 81, pp. 416–423, 2014.

[CAP 96] CAPROS P., Energy efficiency and conversion decentralization: Evidence from the PRIMES model, National Technical University of Athens, 1996.

[CAP 12] CAPROS P., The PRIMES energy system model, Athens, 2012.

[CAR 97] CARRARO C., GALEOTTI M., "Economic growth, international competitiveness and environmental protection: R&D and innovation strategies with the WARM model", *Energy Economics*, vol. 19, no. 1, pp. 2–28, 1997.

[CAS 04] CASTRONUOVO E.D., LOPES J.A.P., "On the optimization of the daily operation of a wind-hydro power plant", *IEEE Transactions on Power Systems*, vol. 19, no. 3, pp. 1599–1606, 2004.

[CER 16] CERRI D., TERZI S., "Proposal of a toolset for the improvement of industrial systems' lifecycle sustainability through the utilization of ICT technologies", *Computers in Industry*, vol. 81, pp. 47–54, 2016.

[CHA 09] CHANDEL M., WILLIAMS E., Synthetic natural gas (SNG): technology, environmental implications, and economics, Climate Change Policy Partnership Duke University, available at: http://www.canadiancleanpowercoalition.com/pdf/SNG3%20-%20synthetic.gas.pdf, January 2009.

[CHA 12] CHAZAN M. *et al.*, "The Oldowan horizon in Wonderwerk Cave (South Africa): archaeological, geological, paleontological and paleoclimatic evidence", *Journal of Human Evolution*, vol. 63, no. 6, pp. 859–866, 2012.

[CHE 13] CHESTER L., "Dissecting the conjunction of capitalism's environmental, energy, and economic crises: the example of one liberal, market-based economy", *Journal of Economic Issues*, vol. XLVII, no. 2, pp. 485–494, 2013.

[CHE 16] CHEN Y., ZHANG T., YANG H. *et al.*, "Study on energy and economic benefits of converting a combined heating and power system to a tri-generation system for sewage treatment plants in subtropical area", *Applied Thermal Engineering*, vol. 94, pp. 24–39, 2016.

[CHO 09] CHO H. *et al.*, "Evaluation of CCHP systems performance based on operational cost, primary energy consumption, and carbon dioxide emission by utilizing an optimal operation scheme", *Applied Energy*, vol. 86, pp. 2540–2549, 2009.

[CHO 11] CHOUCRI N., GOLDSMITH D., TOUFIE M., "Modeling renewable energy readiness: the UAE context", *MIT Openaccess Articles*, IEEE, pp. 211–216, available at:http://hdl.handle.net/1721.1/71264, 2011.

[CHO 16] CHONG W., SHARP S., China's energy landscape, Clyde & Co Insight, available at: http://www.clydeco.com/insight/article/chinas-energy-landscape, 2016.

[CIN 10a] CINAR D., KAYAKUTLU G., "Scenario analysis using Bayesian networks: a case study in energy sector", *Knowledge-Based Systems*, vol. 23, pp. 267–276, 2010.

[CIN 10b] CINAR D., KAYAKUTLU G., DAIM T., "Development of future energy scenarios with intelligent algorithms: case of hydro in Turkey", *Energy*, vol. 35, pp. 1724–1729, 2010.

[CLA 14] CLARK W.W., COOKE G., "The green industrial revolution", in CLARK W. (ed), *Global Sustainable Communities Handbook: Green Design Technologies and Economics*, Butterworth-Heinemann, pp. 13–39, 2014.

[COM 12] COMODI G., CIOCCOLANTI L., GARGIULO M., "Municipal scale scenario: analysis of an Italian seaside town with MarkAL-TIMES", *Energy Policy*, vol. 41, pp. 303–315, 2012.

[COP 15] COP, Sustainable Innovation Forum, available at: http://www.cop21paris.org/, accessed on December 2015.

[COS 16] COSYTEC, available at: http://www.cosytec.com, accessed on May 2015.

[CÔT 95] CÔTÉ R., HALL J., "Industrial parks as ecosystems", *Journal of Cleaner Production*, vol. 3, nos 1–2, pp. 41–46, 1995.

[COU 87] COUNCIL on FOREIGN RELATIONS (CFR), Report of the World Commission on Environment and Development: Our Common Future (Brundtland Report), available at: http://www.cfr.org/economic-development/report-world-commission-environment-development-our-common-future-brundtland-report/p26349, March 1987.

[COW 10] COWAN D.M. *et al.*, "A cross-sectional analysis of reported corporate environmental sustainability practices", *Regulatory Toxicology and Pharmacology*, vol. 58, no. 3, pp. 524–538, 2010.

[COW 15] COWELL R. *et al.*, "Rescaling the governance of renewable energy: lessons from the UK devolution experience", *Journal of Environmental Policy & Planning*, vol. 7200, pp. 1–23, June 2015.

[CRA 92] CRAS J.-Y., *A Review of Industrial Constraint Solving Tools*, AI Intelligence, AI Watch, Oxford, UK, 1992.

[DAG 00] DAG S., Volvo faces a deregulated European electricity market, Linköping University, Sweden, 2000.

[DAI 09] DAIAN G., OZARSKA B., "Wood waste management practices and strategies to increase sustainability standards in the Australian wooden furniture", *Journal of Cleaner Production*, vol. 17, no. 17, pp. 1594–1602, 2009.

[DAI 10] DAIM T.U., KAYAKUTLU G., COWAN K., "Developing Oregon's renewable energy portfolio using fuzzy goal programming model", *Computers and Industrial Engineering*, vol. 59, pp. 786–793, 2010.

[DAI 13] DAIM T.U., KAYAKUTLU G., SUHARTO Y. *et al.*, "Clean energy investment scenarios using the Bayesian network", *International Journal of Sustainable Energy*, vol. 32, no. 4, pp. 1–12, 2013.

[DAL 15] DALTON M., STEINHAUSER G., "Nations unite in global agreement on elimate change", *Wall Street Journal*, Available at: http://www.wsj.com/articles/final-draft-of-global-climate-change-deal-is-complete-1449906731, 2015.

[DE 80] DE BONO E., *Future Positive*, Penguin Books, 1980.

[DEN 59] DENNIS J.B., *Mathematical Programming and Electrical Networks,* Wiley, New York, 1959.

[DEN 83] DENNIS J.E., SCHNABEL R.B., *Numerical Methods for Unconstrained Optimization and Nonlinear Equations*, Prentice-Hall, Inc., Englewood Cliffs, 1983.

[DEN 15] DENG Q., JIANG X., ZHANG L.A. *et al.*, "Making optimal investment decisions for energy service companies under uncertainty: a case study", *Energy*, vol. 88, pp. 234–243, 2015.

[DER 13] DEREK O., BIGG T., *Earth Summit II: Outcomes and Analysis*, Routledge, 2013.

[DIA 06] DIAS R.A., MATTOS C.R., BALESTIERI J.A.P., "The limits of human development and the use of energy and natural resources", *Energy Policy*, vol. 34, pp. 1026–1031, 2006.

[DIB 96] DIBONA D.K., *Global Warming, Energy Efficiency and the Role of the Built Environment Download*, MIT, available at: http://hdl.handle.net/1721.1/58655, 1996.

[DIN 88] DINCBAS M., VAN HENTENRYNCK P., SIMONIS H. *et al.*, "The CHIP system: constraint handling", *Prolog of CADE (International Conference on automatic Deduction)*, Argone, 1988.

[DIO 15] DIOUF B., PODE R., OSEI R., "Recycling mobile phone batteries for lighting", *Renewable Energy*, vol. 78, pp. 509–515, 2015.

[DOE 16] DOE, "Renewable energy technologies", available at: http://energy.gov/eere/wind/animation-how-wind-turbine-works, accessed on 25 June 2016.

[DOL 96] DOLENC N., LIBRALESSO J.M., THIRION C. *et al.*, "The SACHEM Project: blast furnace operating support system; ambition and stakes, development and first results", *3rd European Ironmaking Congress*, Ghent, Belgium, 16–18 September 1996.

[DOM 11] DOMÍNGUEZ M. *et al.*, "Analysis of electric power consumption using self-organizing maps", *Proceedings of the 18th IFAC World Congress*, pp. 12213–12218, 2011.

[DOR 99] DORIGO M., DI CARO G., "The ant colony optimization meta-heuristic", *New Ideas in Optimization*, vol. 2, pp. 11–32, available at: http://portal.acm.org/citation.cfm?id=329055.329062, 1999.

[DOR 06] DORIGO M., BIRATTARI M., STÜTZLE T., "Ant colony optimization", *IEEE Computational Intelligence Magazine*, vol. 1, no. 4, pp. 28–39, available at: http://ieeexplore.ieee.org/lpdocs/epic03/wrapper.htm?arnumber=4129846, 2006.

[DU 15] DU FLOCON A LA VAGUE, available at: http://www.dufloconalavague.org/, 2015.

[EDF 15] EDF, Des drones au service du réseau numérique, available at: http://alsace.edf.com/innover/des-drones-au-service-du-réseau-électrique/, accessed 2015.

[ELH 07] EL-HAGGAR D.S.M., *Sustainable Industrial Design and Waste Management*, available at: http://www.sciencedirect.com/science/article/pii/B9780123736239500101, Elsevier, 2007.

[EMP 16] EMPRENTE H20, available at: http://www.empreinteh2o.com/, 2016.

[ENA 10] ENAMI N. *et al.*, "Neural network based energy efficiency in wireless sensor networks: a survey", *International Journal of Computer Science & Engineering Survey (IJCSES)*, vol. 1, no. 1, pp. 39–55, 2010.

[ENE 06] ENERGY POLICIES OF IEA COUNTRIES, Energy review, p. 172, 2006.

[ENE 11] ENERCITIES, available at: http://www.enercities.eu/game.php, 2011.

[ENE 16] ENERGY STORAGE ASSOCIATION, accessed June 2016, available at: http://energystorage.org/compressed-air-energy-storage-caes.

[ENT 03] ENTOVATION, available at: http://www.entovation.com/group-alliance/en2polis.htm.

[ENV 16] ENVIRONMENT GURU, available at: http://www.environmentguru.com/pages/elements/element.aspx?id=3883644#ixzz4HCSPNQEp, accessed August 2016.

[ERC 14] ERCAN S., KAYAKUTLU G., "Patent value analysis using support vector machines", *Soft Computing*, vol. 2013, no. 1, pp. 1–16, 2014.

[ERC 15] ERCAN, S., KAYAKUTLU G., "Scheduling in energy systems", *SIGMA*, vol. 33, no. 4, pp. 679–690, 2015.

[ERN 78] ERNST R.B., *Aggregate Energy, Efficiency and Productivity*, MIT Publishing, 1978.

[EUR 95] EUROPEAN COMMISSION, "W.A.R.M. world assessment of resource management", European Commission, Technical Report Version 1.5 – October 1995, available at: bookshop.europa.eu/bg/warm.../CGNA16716ENC_001.pdf, 1995.

[EUR 06] EUROPE INNOVA THEMATIC WORKSHOP, "Lead markets and innovation", Summary Report, Munich, Germany, 2006.

[EUR 14] EUROPEAN COMMISSION, 2030 Energy strategy, available at: http://ec.europa.eu/energy/en/topics/energy-strategy/2030-energy-strategy, 2014.

[EUR 15] EUROPEAN COMMISSION, Milesecure-2050, European Commision Policy Brief, available at: https://ec.europa.eu/research/social-sciences/pdf/policy_briefs/milesecure2050_policy-brief_032015.pdf, 2015.

[EUR 16b] European Parliament Policy Department "Industry 4.0", available at: http://www.europarl.europa.eu/RegData/etudes/STUD/2016/570007/IPOL_STU(2016)570007_EN.pdf, 2016.

[EUR 16c] European Research Results, available at: http://bookshop.europa.eu/en/research-eu-results-magazine-pbZZAC14008/, 2016.

[FAR 16a] FARID M., KEEN M., PAPAIOANNOU M. *et al.*, *After Paris: Fiscal, Macroeconomic, and Financial Implications of Climate Change,* IMF Staff Discussion Note, available at: https://www.imf.org/external/pubs/ft/sdn/2016/sdn1601.pdf, 2016.

[FAR 16b] FARLEY J., MALGHAN D. (eds), *Beyond Uneconomic Growth Economics, Equity and the Ecological Predicament*, Edward Edgar Publishing, 2016.

[FAZ 14] FAZLOLLAHI S., BECKER G., MARÉCHAL F., "Multi-objectives, multi-period optimization of district energy systems: III. Distribution networks", *Computers & Chemical Engineering*, vol. 66, pp. 82–97, 2014.

[FED 14] FEDENKO V.S., SHEMET S.A., Method for assessment of plant resistance to metals, Patent UA 91156U, 2014.

[FEI 14] FEI Q., "The impacts of energy prices and technological innovation on the fossil fuel-related electricity-growth nexus: an assessment of four net energy exporting countries", *Journal of Energy in Southern Africa*, vol. 25, no. 3, pp. 46–60, 2014.

[FEL 16] FELDMAN V., "5 key trends for the future of clean energy", World Economic Forum News Letter, available at: https://www.weforum.org/agenda/2016/01/5-key-trends-for-the-future-of-clean-energy/, 2016.

[FEN 13] FENG Z., RIO M., ALLAIS R. *et al.*, "Toward a systemic navigation framework to integrate sustainable development into the company", *Journal of Cleaner Production*, vol. 54, pp. 199–214, 2013.

[FER 95] FERBER J., *Les systèmes multi-agents: vers une intelligence collective*, Interéditions, Paris, 1995.

[FIT 15] FITIWI D.Z. *et al.*, "A new approach of clustering operational states for power network expansion planning problems dealing with RES (renewable energy source) generation operational variability and uncertainty", *Energy*, vol. 90, pp. 1360–1376, 2015.

[FON 10] FONG K.F. *et al.*, "Energy management and design of centralized air-conditioning systems through the non-revisiting strategy for heuristic optimization methods", *Applied Energy*, vol. 87, no. 11, pp. 3494–3506, 2010.

[FOU 12] FOUQUET R., PEARSON P.J.G., "Past and prospective energy transitions: insights from history", *Energy Policy*, vol. 50, pp. 1–7, 2012.

[GAN 02] GANAPATHY V., "Heat recovery boilers", Chapter in *Industrial Boilers and Heat Recovery Steam Generators Design, Applications, and Calculations*, CRC Press, 2002.

[GAR 12] GARETTI M., TAISCH M., "Sustainable manufacturing: trends and research challenges", *Production Planning & Control*, vol. 23, nos 2–3, pp. 83–104, 2012.

[GAS 01] GASS S.I., HARRIS C.M. (eds), *Encyclopedia of Operations Research and Management Science*, Kluwer Academic Publishers, 2001.

[GDF 16] GDF ENERGY "Corporate Social Responsibility" available at: http://www.gdfsuezau.com/about-us/Corporate-Social-Responsibility, accessed on August 2016.

[GEI 07] GEIDL M. *et al.*, "Energy hubs for the future", *IEEE Power and Energy Magazine*, vol. 5, no. 1, pp. 24–30, 2007.

[GEN 15] GENERAL ASSEMBLY OF UNITED NATIONS, Transforming our world: The 2030 agenda for sustainable development, UNDPI 2015.

[GER 16] GERSTLBERGER W., KNUDSEN M.B., DACHS B. *et al.*, "Closing the energy-efficiency technology gap in European firms? Innovation and adoption of energy efficiency technologies", *Journal of Engineering and Technology Management*, vol. 40, pp. 87–100, 2016.

[GHA 13] GHANBARI A. *et al.*, "A cooperative ant colony optimization-genetic algorithm approach for construction of energy demand forecasting knowledge-based expert systems", *Knowledge-Based Systems*, vol. 39, pp. 194–206, available at: http://dx.doi.org/10.1016/j.knosys.2012.10.017, 2013.

[GIB 96] GIBBONS L.E., HEARN D.W., PARDALOS P.M., "A continuous based heuristic for the maximum clique problem", in D.S. JOHNSON, M.A. TRICK (eds): *Cliques, Coloring, and Satisfiability: Second DIMACS Implementation Challenge*, American Mathematical Society, pp. 103–124, 1996.

[GIF 07] GIFFINGER R., "Smart cities. Ranking of European medium-sized cities", available at: http://www.smart-cities.eu/download/smart_cities_final_report.pdf, October 2007.

[GIR 15] GIRONES V.C. *et al.*, "Strategic energy planning for large-scale energy systems: a modelling framework to aid decision-making", *Energy*, vol. 90, Part 1, pp. 173–186, 2015.

[GLA 15] GLASNOVIC Z., MARGETA K., PREMEC K., "Could Key Engine, as a new open-source for RES technology development, start the third industrial revolution?", *Renewable and Sustainable Energy Reviews*, vol. 57, pp. 1194–1209, 2015.

[GOL 89] GOLDBERG D., *Genetic Algorithms in Search, Optimization and Machine Learning*, Addison-Wesley, 1989.

[GOL 16] GOLDMAN D., "Human-carrying drone debuts at CES", CNN Money, available at: http://money.cnn.com/2016/01/06/technology/ces-2016-ehang-drone/, 19 January 2016.

[GRA 09] GRANADE H.C. *et al.*, "Unlocking energy efficiency in the US economy", *McKinsey Global Energy and Materials*, pp. 1–165, available at: http://www.mckinsey.com/Client_Service/Electric_Power_and_Natural_Gas/Latest_thinking/Unlocking_energy_efficiency_in_the_US_economy, 2009.

[GRI 97] GRITSEVICH I., DASHEVSKY Y., ZHUZE V., "Prospects for western technologies in improving district heating energy efficiency in Russia", *International Conference on Technologies for Activities*, Vancouver, pp. 655–663, 1997.

[GRO 15] GROSSBERG F., WOLFSON M., MAZUR-STOMMEN S. *et al.*, Gamified energy efficiency programs, Report B1501, American Council for an Energy-Efficient Economy, 2015.

[GRU 99] GRUBLER A., NAKICENOVIC N., VICTOR D.G., "Dynamics of energy technologies and global change", *Energy Policy*, vol. 27, no. 5, pp. 247–280, 1999.

[GRU 08] GRUBER E., PLESSER S., SOFRONIS I. *et al.*, Monitoring electricity consumption in the tertiary sector, EL-TERTIARY, p. 126, 2008.

[GUA 14] GUALBERTI G., BAZILIAN M., MOSS T., "Energy investments in Africa by the US, Europe and China", *International Association for Energy Economics (IAEE) Newsletter*, 2014.

[GUR 11] GURESEN E., KAYAKUTLU G., "Definition of artificial neural networks with comparison to other networks", *Procedia Computer Science*, vol. 3, pp. 426–433, 2011.

[GUE 13] GUEST POST: "History of home heating: from fire in a cave to an electric fireplace", www.hankeringforhistory.com/history-of-home-heating-from-fire-in-a-cave-to-an-electric-fireplace, October 9, 2013.

[GWE 13] GWEC, "Global wind report annual market update", *GWEC*, p. 80, available at: http://www.gwec.net/publications/global-wind-report-2/global-wind-report-2013/, 2013.

[HAL 08] HALL P.J., BAIN E.J., "Energy-storage technologies and electricity generation", *Energy Policy*, vol. 36, no. 12, pp. 4352–4355, 2008.

[HAM 07] HAMMOND G.P., "Industrial energy analysis, thermodynamics and sustainability", *Applied Energy*, vol. 84, pp. 675–700, 2007.

[HAR 16] HARDTALK, "Fatih Birol at Hardtalk", available at www.bbc.co.uk/programmes/b07chs3d, 2016.

[HAS 06] HASAN A.T. *et al.*, "An adaptive-learning algorithm to solve the inverse kinematics problem of a 6 D.O.F serial robot manipulator", *Advances in Engineering Software*, vol. 37, pp. 432–438, 2006.

[HAT 05] HATLEY D.D., MEADOR R J, KATIPAMULA S. *et al.*, "Energy management and control system: desired capabilities and functionality", Report No PNNL-15074, Pacific Northwest National Laboratory, U.S. Department of Energy, 2005.

[HEL 90] HELM J.L. (ed.), *Energy: Production, Consumption, and Consequences*, National Academy of Engineering, 1990.

[HER 07] HERRING H., ROY R., "Technological innovation, energy efficient design and the rebound effect", *Technovation*, vol. 27, no. 4, pp. 194–203, 2007.

[HER 09] HERRMANN C., THIEDE S., "Process chain simulation to foster energy efficiency in manufacturing", *CIRP Journal of Manufacturing Science and Technology*, vol. 1, no. 4, pp. 221–229, 2009.

[HER 10] HERNANDEZ P., KENNY P., "From net energy to zero energy buildings: defining life cycle zero energy buildings (LC-ZEB)", *Energy and Buildings*, vol. 42, pp. 815–821, 2010.

[HEW 99] HEWSON M.J.C., "Success with energy management", *Ecotextile*, vol. 98, pp. 33–41, 1999.

[HIS 13] History of Home Heating, available at: http://www.coolinfographics.com/blog/2013/3/20/the-history-of-home-heating.html, 2013.

[HOF 12] HOFFMAN J., Potential health and environmental effects of hydrofracking in the Williston Basin, Montana, 2012, available at: http://serc.carleton.edu/NAGTWorkshops/health/case_studies/hydrofracking_w.html.

[HOL 75] HOLLAND J.H., *Adaptation in Natural and Artificial Systems*, The University of Michigan Press, Ann Arbor, 1975.

[HOO 05] HOOSHMAND R., "Application of a genetic algorithm in the reduction of loss and control of the voltage of distribution networks using the real optimization of fixed and switched capacitors", *European Transactions on Electrical Power*, vol. 15 pp. 425–442, 2005.

[HOR 15] HORTON A., "How the energy landscape in India is quickly changing", *The Good Men Project*, KPMG report, December 13, 2015.

[HOS 90] HOSKINS JOSIAH C., KALLIYUR K.M., HIMMELBLAU DAVID M., "Incipient fault detection and diagnosis using artificial neural networks", *Neural Networks 1990 IJCNN International Joint Conference on*, San Diego, vol. 1, pp. 81–86, 1990.

[HSE 16] HSE, Hydrogen Storage Engineering-Models, available at: http://www.hsecoe.org/models.php, accessed in June 2016.

[HU 15] HU J., Combustion gas turbine for use in automobile industry, has air compressor, electromagnetic valve, fuel mixed gas inlet pipe, combustion chamber, power turbine and exhaust pipe for forming opened combustion-gas circulation, WO2015180540-A1;CN105157060-A, 2015.

[IBM 16] IBM TRIRIGA PROJECT, available at: http://www-03.ibm.com/software/products/en/tririga-energy-optimization, 2016.

[IEA 14a] IEA, Key World Energy Statistics, International Energy Agency report, OECD/IEA, 2014.

[IEA 14b] IEA, Key World Energy Outlook, International Energy Agency report, OECD/IEA, 2014.

[IEA 14c] IEA, Linking heat and electricity systems, International Energy Agency report OECD/IEA, 2014.

[IEA 15] IEA, World Energy Outlook, International Energy Agency report, OECD/IEA, 2015.

[IEA 15] Energy and Climate Change, World Energy Outlook, Special Briefing for COP21, OCDE/IEA, 2015.

[IEE 16] IEEE Power & Energy Society Multi-agent Systems Working Group, available at: http://sites.ieee.org/pes-mas/, accessed 2016.

[ISS 15] IISY GRID, First operational district level smart grid in France, available at: http://www.issy.com/home/issy-a-smart-city/issygrid, 2015.

[IMR 14] IMRAN A.M., KOWSALYA M., "A new power system reconfiguration scheme for power loss minimization and voltage profile enhancement using fireworks algorithm", *International Journal of Electrical Power and Energy Systems*, vol. 62, pp. 312–322, 2014.

[INT 02] INTERNATIONAL ENERGY AGENCY, World Energy Outlook, 2002.

[INT 10] INTERNATIONAL ENERGY AGENCY, Energy technology perspectives: Scenarios & strategies to 2050, available at: http://www.oecd-ilibrary.org.ezproxy.library.uq.edu.au/energy/energy-technology-perspectives-2010_energy_tech-2010-en , 2010.

[INT 12] INTERNATIONAL ENERGY AGENCY "Understanding energy challenges in India", *IEEE Industry Applications Magazine*, vol. 17, no. 2, pp. 32–37, 2012.

[INT 13] INTELLIGENT ENERGY EUROPE, 2020 Energy, supported by Intelligent Energy Europe, available at: http://www.2020energy.eu/en, accessed on 26 July 2013.

[IRA 08] IRA–INSTITUT DE REGULATION ET D'AUTOMATION, available at: http://www.ira.eu/institut_regulation_automation.php, 2008.

[IWA 02] IWAMATSU M., "Co-evolutionary global optimization algorithm", *Proceedings of the 2002 Congress on Evolutionary Computation CEC02 Cat No02TH8600*, pp. 0–4, 2002.

[JAC 01] JACOBSEN H.K "Technological progress and long-term energy demand – a survey of recent approaches and a Danish case", *Energy Policy*, vol. 29, pp. 147–157, 2001.

[JAY 10] JAYAL A.D. *et al.*, "Sustainable manufacturing: modeling and optimization challenges at the product, process and system levels", *CIRP Journal of Manufacturing Science and Technology*, vol. 2, no. 3, pp. 144–152, 2010.

[JEV 66] JEVONS W.S., *The Coal Question; An Inquiry Concerning the Progress of the Nation, and the Probable Exhaustion of Our Coal Mines*, 2nd ed., McMillian & Co, 1866.

[JIA 03] JIA N.X., YOKOYAMA R., "Profit allocation of independent power producers based on cooperative Game theory", *International Journal of Electrical Power and Energy System*, vol. 25, no. 8, pp. 633–641, 2003.

[JIN 12a] JING Y.-Y., BAI H., WANG J.-J., "Multi-objective optimization design and operation strategy analysis of BCHP system based on life cycle assessment", *Energy*, vol. 37, no. 1, pp. 405–416, 2012.

[JIN 12b] JIN T., ZHIGANG T., HUERTA M. *et al.*, "Coordinating maintenance with spares logistics to minimize levelized cost of wind energy", *International Conference on Quality, Reliability, Risk, Maintenance, and Safety Engineering (ICQR2MSE)*, pp. 1022–1027, 2012.

[JOS 03] JOSKOW P.L., Energy policies and their consequences after 25 years, MIT Center for Energy, available at: http://hdl.handle.net/1721.1/45006, 2003.

[JOS 14] JOSANG A., "Identity management and trusted interaction in internet and mobile computing", *IET Information Security*, vol. 8, no. 2, pp. 67–79, 2014.

[JRA 14] JRADI M., RIFFAT S., "Tri-generation systems: energy policies, prime movers, cooling technologies, configurations and operation strategies", *Renewable and Sustainable Energy Reviews*, vol. 32, pp. 396–415, 2014.

[JRI 16] JRIDI O. *et al.*, "Energy paradox and political intervention: a stochastic model for the case of electrical equipments", *Energy Policy*, vol. 93, pp. 59–69, 2016.

[KAL 00] KALITVENTZEFF B., MARÉCHAL F., "Optimal insertion of energy saving technologies in industrial processes: a web-based tool helps in developments and co-ordination of a European R&D project", *Applied Thermal Engineering*, vol. 20, no. 15, pp. 1347–1364, 2000.

[KAN 10] KANALA R., Social MARKAL: incorporating behavioral parameters, IEA-ETSAP, 2010.

[KAR 10] KARLSSON-VINKHUYZEN S.I., "The United Nations and global energy governance: past challenges, future choices", *Global Change, Peace & Security*, vol. 22 no. 2, pp. 175–195, 2010.

[KAU 09] KAUCIC M., "Predicting EU energy industry excess returns on EU market index via a constrained genetic algorithm", *Computational Economics*, vol. 34, no. 2, pp. 173–193, 2009.

[KAY 15] KAYAKUTLU G., ERCAN S., "Regional energy portfolio construction: case studies in Turkey", in CUCCHIELLA F., KOH S.C.L. (eds), *Sustainable Future Energy Technology and Supply Chains: A Multi-perspective Analysis*, Springer, pp. 107–126, 2015.

[KEE 10] KEEN S., Debunking Economics Model, 2010, available at: http://www.debunkingeconomics.com/.

[KEL 06] KELLY, N., "The role of energy efficiency in reducing Scottish and UK CO_2 emissions", *Energy Policy*, vol. 34, no. 18, pp. 3505–3515, 2006.

[KEN 95] KENNEDY J., EBERHART R., "Particle swarm optimization", *Proceedings of the IEEE International Conference on Neural Networks*, vol. 4, pp. 1942–1948, 1995.

[KER 15] KERN E. *et al.*, "Impacts of software and its engineering on the carbon footprint of ICT", *Environmental Impact Assessment Review*, vol. 52, pp. 53–61, 2015.

[KES 12] KESKIN C., KAYAKUTLU G., "Value stream maps for industrial energy efficiency", *Portland International Center for Management of Engineering and Technology: Technology Management for Emerging Technologies,* PICMET'12, pp. 2824–2831, 2012.

[KIR 01] KIRANOUDIS C.T., VOROS N.G., MAROULIS Z.B., "Short-cut design of wind farms", *Energy Policy*, vol. 29, no. 7, pp. 567–578, 2001.

[KIT 06] KITAMURA S. *et al.*, "Modified multiobjective particle swarm optimization method and its application to energy management system for factories", *Electrical Engineering in Japan*, vol. 156, no. 4, pp. 33–42, 2006.

[KNA 15] KNAPTON S., "Power plants: how scientists could harness energy from flowers", available at: http://www.telegraph.co.uk/news/science/12008797/Power-plants-how-scientists-could-harness-energy-from-flowers.html, 2015.

[KOH 90] KOHONEN T., "The self-organizing map", *Proceedings of the IEEE*, vol. 78, no. 9, pp. 1464–1480, 1990.

[KOR 01] KORHONEN J., "Regional industrial ecosystems – extended", *Eco-Management and Auditing*, vol. 69, pp. 57–69, 2001.

[KRA 10] KRAMER R.A., BRANCH R., LIU W., "Optimizing building energy efficiency with modeling and intelligent controls", *Energy Engineering*, vol. 107, no. 2, pp. 52–80, 2010.

[KUM 78] KUMMEL R., GROSCURTH H.M., VAN GOOL W., Energy optimization in industrial models, Energy Science Project Report, State University of Utrecht, Holland, 1978.

[KUM 13] KUMAR S., CHATURVEDI D.K., "Optimal power flow solution using fuzzy evolutionary and swarm optimization", *International Journal of Electrical Power & Energy Systems*, vol. 47, pp. 416–423, 2013.

[LAG 15] LAGRANGE J.-L., *Mécanique Analytique*, Gauthier-Villars et fils, Paris, 1815.

[LAM 10] LAM H.L., VARBANOV P.S., KLEMEŠ J.J., "Optimisation of regional energy supply chains utilising renewables: P-graph approach", *Computers and Chemical Engineering*, vol. 34, no. 5, pp. 782–792, 2010.

[LAR 15] LARGILLIER R., SCHNEIDER Corporate presentation, available at: http://www2.schneider-electric.com/documents/events/WFES_2012_Schneider-Electric-and-Smart-Cities.pdf, 2015.

[LAU 12] LAUVERGEON A., Un principe et sept ambitions pour l'innovation – Report, available at: http://www.direccte.gouv.fr/IMG/pdf/LAUVERGEON_1_principe_et_7_ambitions.pdf, 2012.

[LEE 10] LEE K.-H. *et al.*, "Numerical optimization of site selection for offshore wind turbine installation using genetic algorithm", *Current Applied Physics*, vol. 10, no. 2, pp. S302–S306, 2010.

[LEE 15] LEE J., BAGHERI B., KAO H.A., "A cyber-physical systems architecture for Industry 4.0-based manufacturing systems", *Manufacturing Letters*, vol. 3, pp. 18–23, 2015.

[LEI 16] LEITHAO P., COLOMBO A.W., KARNOUSKOS S., "Industrial automation based on cyber-physical systems technologies: prototype implementations and challenges", *Computers in Industry*, vol. 81, pp. 11–25, 2016.

[LEP 14] LEPKOVA N., ZUBKA D., JENSEN R.L., *Global Sustainable Communities Handbook*, Butterworth-Heinemann, Available at: http://www.sciencedirect.com/science/article/pii/B9780123979148000102, 2014.

[LES 16] LESSER R., REEVES M., HARNESS J., "Saving globalization and technology from themselves: imperatives for corporate leaders", BCG Perspectives, available at: https://www.bcgperspectives.com, 2016.

[LET 08] LETCHER T.M. (ed.), *Future Energy: Improved, Sustainable and Clean Options for our Planet*, Elsevier Ltd., 2008.

[LI 12] LI G., SHI J., "Applications of Bayesian methods in wind energy conversion systems", *Renewable Energy*, vol. 43, pp. 1–8, 2012.

[LIK 03] LIKA K., KOOIJMAN S.A.L.M., "Life history implications of allocation to growth versus reproduction in dynamic energy budgets", *Bulletin of Mathematical Biology*, vol. 65, no. 5, pp. 809–834, 2003.

[LIN 12] LIN B., LIU X., "Dilemma between economic development and energy conservation: Energy rebound effect in China", *Energy*, vol. 45, no. 1, pp. 867–873, 2012.

[LIN 13] LIND A. *et al.*, "Analysis of the EU renewable energy directive by a techno-economic optimisation model", *Energy Policy*, vol. 60, pp. 364–377, 2013.

[LO 16] LO PRETE B., HOBBS C., "A cooperative game theoretic analysis of incentives for microgrids in regulated electricity markets", *Applied Energy*, vol. 169, pp. 524–541, 2016.

[LOU 08] LOULOU R., LABRIET M., "ETSAP-TIAM: the TIMES integrated assessment model Part I: model structure", *Computational Management Science*, vol. 5, nos 1–2, pp. 7–40, 2008.

[LUI 03] LUITEN E., BLOK K., "The success of a simple network in developing an innovative energy-efficient technology", *Energy*, vol. 28, no. 4, pp. 361–391, 2003.

[LUN 16] LUND J.W., BOYD T.L., "Direct utilization of geothermal energy 2015 worldwide review", *Geothermics*, vol. 60, pp. 66–93, 2016.

[MAD 97] MADAN S., BOLLINGER K.E., "Applications of artificial intelligence in power systems", *Electric Power Systems Research*, vol. 41, no. 2, pp. 117–131, 1997.

[MAG 14] MAGGETTI M., "The politics of network governance in Europe: the case of energy regulation", *West European Politics*, vol. 37, no. 3, pp. 497–514, 2014.

[MAL 10] MALANIMA P., "Energy in history", in *Encyclopedia of Life Support Systems*, (UNESCO-Eolss), pp. 1–26, 2010.

[MAN 94] MANAGO M., MERCIER-LAURENT E., Integrating induction and casebased reasoning for diagnostic problems with focus on medical domains, project, www.cordis.europa.eu/project/rcn/33561_en.html, 1994.

[MAR 09] RIEDMILLER M., Albert-Ludwigs, University Freiburg, available at: http://ml.informatik.uni-freiburg.de/_media/documents/teaching/ss09/ml/perceptrons.pdf, 2009.

[MAR 13] MARTINEZ-VAL M.J., "Energy for sustainable development: a systematic approach for a badly defined challenge", *Energy Conversion and Management*, vol. 72, pp. 3–11, 2013.

[MAR 16] MARKAL and TIMES training, available at: http://www.iea-etsap.org, accessed May 2016.

[MAS 06] MASON A., "Images of the energy future", *Environmental Research Letters*, vol. 1, no. 1, p. 014002, 2006.

[MEM 15] MEMARZADEH M., POZZI M., KOLTER J.Z., "Optimal planning and learning in uncertain environments for the management of wind farms", *Journal of Computing in Civil Engineering*, vol. 29, no. 5, pp. 1–10, 2015.

[MER 07] MERCIER-LAURENT E., Rôle de l'ordinateur dans le processus global de l'innovation à partir des connaissances, Université Jean Moulin Lyon 3, 2007.

[MER 11] MERCIER-LAURENT E., *Innovation Ecosystems*, ISTE Ltd., London, and John Wiley & Sons, New York, 2011.

[MER 13] MERCIER-LAURENT E., "Innovating corporate management: introducing environmental aspects to design activities", *Federated Conference on Computer Science and Information Systems,* Krakow, 2013.

[MER 15] MERCIER-LAURENT E., *The Innovation Biosphere: Planet and Brains in the Digital Era*, ISTE Ltd., London and John Wiley & Sons, New York, 2015.

[MER 16] MERCIER-LAURENT E., "AI & Forecasting", Open Innovation 2.0, Amsterdam, available at: https://ec.europa.eu/digital-single-market/en/news/open-innovation-20-conference-2016-it-happened, 2016.

[MIC 81] MICHALSKI R.S., CARBONELL G.J., MITCHELL T.M., *Machine Learning: An Artificial Intelligence Approach,* vols I–II, 1st ed., Morgan Kauffman Publishers, 1981.

[MIC 94] MICHALSKI R.S., *Machine Learning: A Multistrategy Approach,* Morgan Kaufmann Publishers, 1994.

[MIR 16] MIROVA, www.mirova.com, accessed 2016.

[MIT 09] MITCHELL T., "Carbon democracy", *Economy and Society*, vol. 38, no. 3, pp. 399–432, 2009.

[MIT 11] PATEL P., "Powering your car with waste heat", MIT Technology Review, available at: https://www.technologyreview.com/s/424092/powering-your-car-with-waste-heat/, 2011.

[MOH 14] MOHAMADIBAGHMOLAEI M. *et al.*, "Assessing and optimization of pipeline system performance using intelligent systems", *Journal of Natural Gas Science and Engineering*, vol. 18, pp. 64–76, 2014.

[MON 16] MONTGOMERY A., "Fosters and Nissan outline renewable energy future – based on the car", *Design Week*, 2016.

[MOR 14] MORADI M.H., ESKANDARI M., "A hybrid method for simultaneous optimization of DG capacity and operational strategy in microgrids considering uncertainty in electricity price forecasting", *Renewable Energy*, vol. 68, pp. 697–714, 2014.

[MOZ 07] MOZZO M.A., "Waste heat recovery", in CAPEHART B.L. (ed.), *Encyclopedia of Energy Engineering and Technology*, CRC Press, pp. 1536–1540, 2007.

[MUC 09] MUCUK M., UYSAL D., "Türkiye Ekonomisinde Enerji Tüketimi ve Ekonomik Büyüme", *Maliye Dergisi*, vol. July–December, no. 157, pp. 105–115, 2009.

[MUL 15] MULDER P., "International specialization, sector structure and the evolution of manufacturing energy intensity in OECD countries", *Energy Journal*, vol. 36, no. 3, pp. 111–135, 2015.

[MUÑ 09] MUÑOZ J.I. *et al.*, "Optimal investment portfolio in renewable energy: the Spanish case", *Energy Policy*, vol. 37, no. 12, pp. 5273–5284, 2009.

[MUR 15] MURRAY C., "A closer look at NASA's Orion glass cockpit", Design News, available at: http://www.designnews.com/author.asp?section_id=1395&doc_id=278234, 2015.

[NAG 00] NAGEL J., "Determination of an economic energy supply structure based on biomass using a mixed-integer linear optimization model", *Ecological Engineering*, vol. 16, pp. 91–102, 2000.

[NAG 10] NAGENDRA P., DEY S.H., DUTTA T., "Artificial neural network application for power transfer capability and voltage calculations in multi-area power system", *Leonardo Electronic Journal of Practices and Technologies*, vol. 16, pp. 119–128, 2010.

[NAT 09] NATIONAL RESEARCH CENTER, *America's Energy Future: Technology and Transformation*, The National Academies Press, 2009.

[NEW 87] NEWTON I., *Philosophiæ Naturalis Principia Mathematica (Adv.b.39.1)*, 1687.

[NEW 10] NEWBOLD S.C., "Summary of DICE Model", available at: https://yosemite.epa.gov/ee/epa/eerm.nsf/vwan/ee-0564-114.pdf/$file/ee-0564-114.pdf, 2010.

[NGU 11] NGUENE G. *et al.*, "SOCIO-MARKAL: integrating energy consumption behavioral changes in the technological optimization framework", *Energy for Sustainable Development*, vol. 15, no. 1, pp. 73–83, 2011.

[NIE 95] NIELSEN S.N., "Optimization of exergy in a structural dynamic model", *Ecological Modelling*, vol. 77, nos 2–3, pp. 111–122, 1995.

[NOR 95] NORDHAUS W.D., "The ghosts of climates past and the specters of climate change future", *Energy Policy*, vol. 23, nos 4–5, pp. 269–282, 1995.

[NOR 99] NORDHAUS W.D., BOYER J.G., "Requiem for Kyoto: an economic analysis of the Kyoto protocol", *Energy Journal*, vol. 20, pp. 93–130, 1999.

[NØR 12] NØRSTEBØ V.S., MIDTHUN K., BJØRKVOLL T., "Analysis of carbon capture in an industrial park – A case study", *International Journal of Greenhouse Gas Control*, vol. 9, pp. 52–61, 2012.

[OME 98] OMER S., "Design optimization of thermoelectric devices for solar power generation", *Solar Energy Materials and Solar Cells*, vol. 53, nos 1–2, pp. 67–82, 1998.

[OPE 16] OPENEI, MARKet ALlocation available at: http://en.openei.org/wiki/MARKet_ALlocation_(MARKAL), accessed 2016.

[ORD 15] ORDONEZ DE PABLOS P., EDVINSSON L. (eds), *Intellectual Capital in Organizations. Nonfinancial Reports and Accounts*, Rutgers, 2015.

[OSE 13] OSEMOSYS open software, available at: http://www.osemosys.org/resources.html, 2013.

[PAL 92] PALMER G., "The earth summit: what went wrong at Rio?", *Washington University Law Review*, vol. 70, no. 4, available at: http://openscholarship.wustl.edu/law_lawreview/vol70/iss4/1, 1992.

[PAL 13] PALIT D., SARANGI G.K., The Mini-grid Experience from India, Global Network on Energy for Sustainable Development (GNESD), 2013.

[PAP 98] PAPALEXANDRI K.P., PISTIKOPOULOS E.N., KALITVENTZEFF B., "Modelling and optimization aspects in energy management and plant operation with variable energy demands-application to industrial problems", *Computers and Chemical Engineering* vol. 22, no. 9, pp. 1319–1333, 1998.

[PAP 14] PAPADIMITRIOU T., GOGAS P., STATHAKIS E., "Forecasting energy markets using support vector machines", *Energy Economics*, vol. 44, pp. 135–142, 2014.

[PAR 00] PARASURAMAN R., SHERIDAN T.B., WICKENS C.D., "A model for types and levels of human interaction with automation", *IEEE Transactions Systems, Man, and Cybernetics Part A: Systems and Humans*, vol. 30, no. 3, pp. 286–297, 2000.

[PAS 13] PASQUAL C.O.G.: Values, the triple bottom line in the SME's and corporate sector, http://www.eoi.es/blogs/carlosomargarcia/author/carlosomargarcia, 2013.

[PAS 16] PASSIVE HOUSE DAYS, available at: http://www.passivhausprojekte.de/index.php?lang=en, accessed 2016.

[PAT 11] PATRICE N., "Quatre approches de l'innovation", *Documentaliste – Sciences de l'information*, vol. 48, no. 1, pp. 24–27, 2011.

[PEL 12] PELENUR M.J., CRUICKSHANK H.J., "Closing the energy efficiency gap: a study linking demographics with barriers to adopting energy efficiency measures in the home", *Energy*, vol. 47, no. 1, pp. 348–357, 2012.

[PEN 06] PENALVA J.M., *Intelligence Collective*, Mines, Paris, 2006.

[PER 10] PEREIRA R.N., VICENTE A.A., "Environmental impact of novel thermal and non-thermal technologies in food processing", *Food Research International*, vol. 43, no. 7, pp. 1936–1943, 2010.

[PER 93] PERRIN N., SIBLEY R.M., "Dynamic models of energy allocation and investment", *Annual Review of Ecology and Systematics*, vol. 24, pp. 379–410, 1993.

[PIE 13] PIEROBON, L. *et al.*, "Multi-objective optimization of organic Rankine cycles for waste heat recovery: Application in an offshore platform", *Energy*, vol. 58, pp. 538–549, 2013.

[PHA 16] PHASE CHANGING MATERIALS, available at: http://www.climatetechwiki.org/technology/jiqweb-pcm-0, accessed 2016.

[PIE 16] PIESSE M., South Africa: drought threatens food, energy and water security, Strategic Analysis Paper for Future Directions, Australia, available at: http://www.climatetechwiki.org/technology/jiqweb-pcm-0, 2016.

[PLA 15] PLANETSOLAR, available at: http://www.planetsolar.org/, 2015.

[POP 04] POPP D., "ENTICE: endogenous technological change in the DICE model of global warming", *Journal of Environmental Economics and Management*, vol. 48, no. 1, pp. 742–768, 2004.

[PRA 11] PRATHEEPRAJ E., ABRAHAM A., DEEPA S.N. *et al.*, "Very short term wind power forecasting using PSO-neural network hybrid system", *Advances in Computing and Communications*, vol. 192, pp. 503–511, 2011.

[PRA 15a] PRASAD K. "The Indian market for renewable energy: the 100 Gw solar initiative", Presentation of Il&Fs Energy Development Company Ltd, Munich, Germany, 2015.

[PRA 15b] PRABHU V.V., TRENTESAUX D., TAISCH M., "Energy-aware manufacturing operations", *International Journal of Production Research*, vol. 53, no. 23, pp. 6994–7004, 2015.

[PRO 15] PROSKURYAKOVA L., KOVALEV A., "Measuring energy efficiency: is energy intensity a good evidence base?", *Applied Energy*, vol. 138, pp. 450–459, 2015.

[PTA 06] PTASINSKI K.J., KOYMANS M.N., VERSPAGEN H.H.G., "Performance of the Dutch energy sector based on energy, exergy and extended exergy accounting", *Energy*, vol. 31, no. 15, pp. 3135–3144, 2006.

[QUI 83] QUINLAN J.R., MICHALSKI R.S., *Machine Learning – An Artificial Intelligence Approach*, Tioga, Palo Alto, pp. 463–482, 1983.

[RAD 15] RADOVANOVIC M., FILIPOVIC S., "New approach to energy intensity in the EU – total energy and carbon cost approach", *Energy & Environment*, vol. 26, no. 4, pp. 601–616, 2015.

[RAG 15] RAGER M., GAHM C., DENZ F., "Energy-oriented scheduling based on evolutionary algorithms", *Computers & Operations Research*, vol. 54, pp. 218–231, 2015.

[RAI 10] RAINERI R., CONTRERAS G., "Efficient capacity investment and joint production agreements in an oligopolistic electricity market: the HidroAysen joint venture project", *Energy Policy*, vol. 38, no. 11, pp. 6551–6559, 2010.

[RÄS 08] RÄSÄNEN T., RUUSKANEN J., KOLEHMAINEN M., "Reducing energy consumption by using self-organizing maps to create more personalized electricity use information", *Applied Energy*, vol. 85, no. 9, pp. 830–840, 2008.

[RAU 85] RAUCH-HINDIN W.B., HINDIN H.J., *Artificial Intelligence in Business, Science and Industry: Applications*, Prentice-Hall, 1985.

[REN 10] REN H., ZHOUB W., NAKAGAMI K. *et al.*, "Multi-objective optimization for the operation of distributed energy systems considering economic and environmental aspects", *Applied Energy*, vol. 87, no. 12, pp. 3642–3651, 2010.

[REN 13] REN21, Renewables Global Status Report, Available at: www.ren21.net/gsr, 2013.

[REN 15] REN21, Renewables 2015: Global Status Report – key findings, available at: http://link.brightcove.com/services/player/bcpid1722935254001/?autoStart=false&bctid=3603665481001&height=300&secureConnections=true&width=400\nhttp://www.ren21.net/Portals/0/documents/Resources/GSR/2014/GSR2014_KeyFindings_low res.pdf\nhttp://www.ren21.n, 2015.

[REX 15] REXHÄUSER S., LÖSCHEL A., "Invention in energy technologies: comparing energy efficiency and renewable energy inventions at the firm level", *Energy Policy*, vol. 83, pp. 206–217, 2015.

[REZ 12] REZAIE B., ROSEN M., "District heating and cooling: review of technology and potential enhancements", *Applied Energy*, vol. 93, pp. 2–10, 2012.

[RIO 92] RIO DECLARATION, Report of the United Nations Conference on Environment and Development, available at: https://www.un.org/documents/ga/conf151/aconf15126-1annex1.htm, 1992.

[RIO 12] RIO DECLARATION, "United Nations: Rio+20 – The future we want", UN DPI, 2012.

[RIV 06] RIVIERE T., Optimisation des graphes sous contraintes géométriques: création d'un réseau des routes aériennes pour un contrôle sector-less, PhD Thesis, INPT, 2006.

[ROC 16] ROCKY MOUNTAIN INSTITUTE, "The economics of battery energy storage", available at: www.rmi.org, accessed 2016.

[ROO 87] ROOS J.L. (ed.), *Economics and Artificial Intelligence*, Pergamon Press, 1987.

[ROS 95] ROSENTHAL D.H., GRUENSPECHT H.K., "Effects of global warming on energy use for space heating and cooling in the United States", *Energy Journal*, vol. 16, no. 2, p.77, 1995.

[ROS 09] ROSS I.M., *A Primer on Pontryagin's Principle in Optimal Control*, Collegiate Publishers, 2009.

[RUM 08] RUMMER B., "Assessing the cost of fuel reduction treatments: A critical review", *Forest Policy and Economics*, vol. 10, pp. 355–362, 2008.

[RUS 03] RUSSELL S.J., NORVIG P., *Artificial Intelligence: A Modern Approach*, Prentice Hall, 2003.

[SAI 10] SAIDUR R., "A review on electrical motors energy use and energy savings", *Renewable and Sustainable Energy Reviews*, vol. 14, no. 3, pp. 877–898, 2010.

[SAL 98] SALEHFAR H., BENSON S.A., "Electric utility coal quality analysis using artificial neural network techniques", *Neurocomputing*, vol. 23, nos 1–3, pp. 195–206, 1998.

[SAV 96] SAVAGE C.M., *5th Generation Management: Integrating Enterprises through Human Networking*, Butterworth-Heinemann, 1996.

[SAV 10] SAVAGE C.M., "Grokking the multiple interrelated eco-crises", Rev1.6-2010-01-11-EuroNews:Commentvision, http://kee-inc.com/, 2010.

[SAY 15] SAYGIN D. *et al.*, "The implications for renewable energy innovation of doubling the share of renewables in the global energy mix between 2010 and 2030", *Energies*, vol. 8, no. 6, pp. 5828–5865, 2015.

[SCH 12] SCHWARZER S., "Keeping track of our changing environment – from Rio to Rio +20 (1992–2012)", *Environmental Development*, vol. 3, pp. 166–179, 2012.

[SCH 13] SCHIFFNER K.C., "Energy recovery", chapter in *Air Pollution Control Equipment Selection Guide*, CRC Press, pp. 205–212, 2013.

[SCH 14] SCHILLER M., "Hydrogen energy storage: a new solution to the renewable energy intermittency problem", *Renewable Energy World*, available at: http://www.renewableenergyworld.com/articles/2014/07/hydrogen-energy-storage-a-new-solution-to-the-renewable-energy-intermittency-problem.html, pp. 1–9, 2014.

[SCI 96] SCIUBBA E., "A minimum entropy generation procedure for the discrete pseudo-optimization of finned-tube heat exchangers", *Revue Générale de Thermique*, vol. 35, no. 416, pp. 517–525, 1996.

[SCI 12] SCIENCE DAILY, "Butterfly wings 'art of blackness' could boost production of green fuels", *Science Daily*, 26 March 2012.

[SEE 02] SEEBREGTS A.J., GOLDSTEIN G.A., SMEKENS K., "Energy / environmental modeling with the MARKAL family of models", *International Conference on Operations Research 2001*, pp. 75–82, 2002.

[SEK 16] SEKERA J.A., *The Public Economy in Crisis: A Call for a New Public Economics*, Springer, 2016.

[SEN 10] SEN S., "An overview of clean coal technologies I: pre-combustion and post-combustion emission control", *Energy Sources, Part B: Economics, Planning, and Policy*, vol. 5, no. 3, pp. 261–271, 2010.

[SHA 82] SHANK R., *Dynamic Memory: A Theory of Learning in Computers and People*, Cambridge University Press, 1982.

[SHA 11] SHAO S. *et al.*, "Estimation, characteristics, and determinants of energy-related industrial CO2 emissions in Shanghai (China), 1994–2009", *Energy Policy*, vol. 39, pp. 6476–6494, 2011.

[SHA 15] SHARAFI M., ELMEKKAWY T.Y., "Stochastic optimization of hybrid renewable energy systems using sampling average method", *Renewable and Sustainable Energy Reviews*, vol. 52, pp. 1668–1679, 2015.

[SHI 10] SHITTU E., BAKER E., "Optimal energy R&D portfolio investments in response to a carbon tax", *IEEE Transactions on Engineering Management*, vol. 57, no. 4, pp. 547–559, 2010.

[SHO 75] SHORTLIFFE E.H., BUCHANAN B.G., "A model of inexact reasoning in medicine", *Mathematical Biosciences*, vol. 23, nos 3–4, pp. 351–379, 1975.

[SHR 15] SHROUF F., MIRAGLIOTTA G., "Energy management based on internet of things: practices and framework for adoption in production management", *Journal of Cleaner Production*, vol. 100, pp. 235–246, 2015.

[SHT 04] SHTANGEEVA I., AYRAULT S., "Phytoextraction of thorium from soil and water media", *Water, Air and Soil Pollution*, vol. 154, nos. 1–4, pp. 19–35, 2004.

[SMI 94] SMIL V., *Energy in World History*, Westview Press, 1994.

[SOL 15] SOLAR IMPULSE, available at: www.solarimpulse.com, 2015.

[SOL 16] SOLAR THERMAL MAGAZINE, available at: https://solarthermalmagazine.com/2016/08/12/sumitomo-us-hybrid-collaborate-hydrogen-fuel-cell-powered-public-transit/, accessed August 2016.

[SON 15a] SONY Company site, available at: https://www.sonycsl.co.jp/sp/287/, 2015.

[SON 15b] SONY Company site, available at: https://www.sonycsl.co.jp/tag/energy/, 2015.

[SPR 12] SPRENG D., "Global, tackling long-term problems, energy", *Environement & Policy*, vol. 52, p. 336, 2012.

[STA 15] STAVINDROU E., GABRIELSON R., GOMEZ E. *et al.*, *Electronic plants*, available at: dvances.sciencemag.org/content/1/10/e1501136, 2015.

[STA 16] STAUFFER N., "A battery made of molten metals. New battery may offer low-cost, long-lasting storage for the grid", *MIT Energy Initiative*, available at: http://news.mit.edu/2016/battery-molten-metals-0112, 2016.

[STE 82] STERMAN J., Economic Vulnerability and the Energy Transition, MIT, Boston, 1982.

[TAN 12] TANG C.F., TAN E.C., "Electricity consumption and economic growth in portugal: evidence from a multivariate framework analysis", *Energy Journal*, vol. 33, no. 4, pp. 23–48, 2012.

[THO 54] THOMSON W., "On the dynamical theory of heat. Part V. Thermo-electric currents", *Transactions of the Royal Society of Edinburgh*, vol. 21, part I, p. 123, 1854.

[THO 99] THOMPSON B.S., "Environmentally-sensitive design: Leonardo WAS right!", *Materials and Design*, vol. 20, pp. 23–30, 1999.

[THO 09] THOLLANDER P., MARDAN N., KARLSSON M., "Optimization as investment decision support in a Swedish medium-sized iron foundry – a move beyond traditional energy auditing", *Applied Energy*, vol. 86, no. 4, pp. 433–440, 2009.

[THO 14] THORNTON J.C., GONDER J., "System design, analysis, and modeling for hydrogen storage systems", National Renewable Energy Laboratory (NREL) News, available at: https://www.hydrogen.energy.gov/pdfs/review14/st008_thornton_2014_o.pdf, 2014.

[TOK 09] TOKSARI M.D., "Estimating the net electricity energy generation and demand using the ant colony optimization approach: case of Turkey", *Energy Policy*, vol. 37, no. 2009, pp. 1181–1187, 2009.

[TOW 14] TOWLER B.F., *The Futıre of Energy*, Elsevier, 2014.

[TSA 93] TSATSARONIS G., "Thermoeconomic analysis and optimization of energy systems", *Progress in Energy and Combustion Science*, vol. 19, no. 3, pp. 227–257, 1993.

[TUR 50] TURING A., "Computing machinery and intelligence", *Mind*, vol. LIX, no. 236, pp. 433–460, 1950.

[UN 02] UN STATEMENTS: KOFI ANNAN, The Secretary-General Address to the World Summit on Sustainable Development, Johannesburg, available at: http://www.un.org/events/wssd/statements/sgE.htm, 2002.

[ÜNA 15] ÜNAL A.N., ERCAN S., KAYAKUTLU G., "Optimisation studies on tri-generation: a review", *International Journal of Energy Research*, vol. 39, no. 10, pp. 1311–1334, 2015.

[UNI 97] UNITED NATIONS GENERAL ASSEMBLY, Program for the Further Implementation of Agenda 21, available at: http://www.un.org/en/ga/search/view_doc.asp?symbol=A/RES/S-19/2, 1997.

[USE 16] INTERNATIONAL ENERGY OUTLOOK, U.S. Energy Information Agency, http://www.eia.gov/forecasts/ieo/, 2016.

[USH 10] USHER W., STRACHAN N., UK MARKAL modelling – examining decarbonisation pathways in the 2020s on the way to meeting the 2050 emissions, Target final report for the Committee on Climate Change, University College London, available at: https://www.theccc.org.uk/archive/aws2/4th%20Budget/CCC%20MARKAL%20Final%20Report%20-%20UCL%20Nov10.pdf, 2010.

[VEH 05] VEHVILÄINEN I., PYYKKÖNEN T., "Stochastic factor model for electricity spot price – the case of the Nordic market", *Energy Economics*, vol. 27, no. 2, pp. 351–367, 2005.

[VIC 08] VICTOR P.A., "Managing without growth: slower by design not disaster", in *Advances in Ecological Economics*, Edward Elgar Publishing, UK, 2008.

[VLA 15] VLAHOPLUS C., LYONS C., QUINLAN P., "Renewable energy will force change in utilities", *Natural Gas & Electricity*, vol. 31, no. 9, pp. 20–24, available at: http://dx.doi.org/10.1002/gas.21825, 2015.

[WAL 03] WALLACE S.W., FLETEN S.-E., "Stochastic programming models in energy", *Handbooks in Operations Research and Management Science*, vol. 10, pp. 637–677, 2003.

[WAN 09] WANG J., DAI Y., GAO L., "Exergy analyses and parametric optimizations for different cogeneration power plants in cement industry", *Applied Energy*, vol. 86, no. 6, pp. 941–948, 2009.

[WAN 10] WANG Y.W.Y. *et al.*, "Short term load forecasting: a dynamic neural network based genetic algorithm optimization", *Power Electronics and Motion Control Conference (EPE/PEMC), 2010 14th International*, pp. 157–161, 2010.

[WAN 15] WANG X., PALAZOGLU A., EL-FARRA N.H., "Operational optimization and demand response of hybrid renewable energy systems", *Applied Energy*, vol. 143, pp. 324–335, 2015.

[WAN 16] WANG X., CAI H., FLORIG H.K., "Energy-saving implications from supply chain improvement: an exploratory study on China's consumer goods retail system", *Energy Policy*, vol. 95, pp. 411–420, 2016.

[WEB 97] WEBER L., "Some reflections on barriers to the efficient use of energy", *Energy Policy*, vol. 25, no. 10, pp. 833–835, 1997.

[WES 11] WESSNER C., *Building the 21st Century: U.S. China Cooperation on Science, Technology, and Innovations*, The National Academies Press, 2011.

[WHI 90] WHITE J.W.C., "Climate change: the IPCC scientific assessment", Report prepared for IPCC Working Group 1, Intergovernmental Panel on Climate Change, 1990.

[WHI 13] WHITE G., CALLENDER J., Forecasting the US Energy landscape: EY 2013, energy tax and policy survey, Ernst & Young, 2013.

[WIL 04] WILSON T.L., BELL D.G., "Energy conservation and demand control using distribution automation technologies", *Rural Electric Power Conference*, pp. C4–1–12, 2004.

[WIL 16] WILKES UNIVERSITY, "What is hydrofracking?", available at: http://energy.wilkes.edu/pages/156.asp, accessed 2016.

[WIN 16] WIND ENERGY FOUNDATION, "History of Wind Energy" available at: http://windenergyfoundation.org/about-wind-energy/history/, accessed 2016.

[WIR 05] WIRTH J.M., Schelling now: contemporary readings, available at: http://www.loc.gov/catdir/toc/ecip0416/2004006420.html, 2005.

[WIS 16] WISE CITY PARTNERS, available at: http://www.wisecity.hk/partner, accessed 2016.

[WOL 98] WOLFF A., GROEBEL M.J., JANOWSKY R., "SitEModellingTM: A powerful tool for total site energy optimization", *Computers & Chemical Engineering*, vol. 22, (Supplement 1), pp. S1073–S1084, 1998.

[WOO 16] WOOLF N., "Hyperloop One tests high-speed transport propulsion system", available at: https://www.theguardian.com/technology/2016/may/11/hyperloop-supersonic-train-test-propulsion-system, 12 May 2016.

[WOR 03] WORRELL E. *et al.*, "Productivity benefits of industrial energy efficiency measures", *Energy*, vol. 28, no. 11, pp. 1081–1098, 2003.

[WOZ 06] WOZNIAK L., KRUPA J., GRZESIK J., *Innowacje ekologiczne w rozwoju społeczno-gospodarczym*, Wydawnictwo Wyższej Szkoły Informatyki i Zarządzania, Rzeszow, 2006.

[YAN 15] YAN H. *et al.*, "Future evolution of automated demand response system in smart grid for low-carbon economy", *Journal of Modern Power Systems and Clean Energy*, vol. 3, no. 1, pp. 72–81, 2015.

[YU 15] YU W. *et al.*, "Application of multi-objective genetic algorithm to optimize energy efficiency and thermal comfort in building design", *Energy and Buildings*, vol. 88, pp. 135–143, 2015.

[YUN 11] YUN J., PRICE L., "Voluntary energy efficiency agreements in china: history, impact, and future", *European Council for an Energy-Efficient Economy*, vol. 2011, no. 18, pp. 619–627, 2011.

[ZAD 65] ZADEH L.A., "Fuzzy sets", *Information and Control*, vol. 8, pp. 338–353, 1965.

[ZAF 07] ZAFRA-CABEZA A. *et al.*, "Managing risk in semiconductor manufacturing: a stochastic predictive control approach", *Control Engineering Practice*, vol. 15, no. 8, pp. 969–984, 2007.

[ZHA 00] ZHANG G., HUA B., CHEN Q., "Exergoeconomic methodology for analysis and optimization of process systems", *Computers & Chemical Engineering*, vol. 24, nos 2–7, pp. 613–618, 2000.

[ZHA 08] ZHANG J., WANG G., "Energy saving technologies and productive efficiency in the Chinese iron and steel sector", *Energy*, vol. 33, no. 4, pp. 525–537, 2008.

[ZHA 14] ZHAO Y. *et al.*, "A comparative study of energy consumption and efficiency of Japanese and Chinese manufacturing industry", *Energy Policy*, vol. 70, pp. 45–56, 2014.

[ZHU 00] ZHU F.X., VAIDEESWARAN L., "Recent research development of process integration in analysis and optimisation of energy systems", *Applied Thermal Engineering*, vol. 20, nos 15–16, pp. 1381–1392, 2000.

[ZHU 12] ZHU Y. *et al.*, "A novel maximum power point tracking control for permanent magnet direct drive wind energy conversion systems", *Energies*, vol. 5, no. 12, pp. 1398–1412, 2012.

[ZWI 90] ZWINGELSTEIN G., LALEUF J.C., "Application of expert systems to EDF power plants", *Proceedings of the First International Conference on Supercomputing in Nuclear Applications (SNA '90)*, 1990.

Index

G, H, I

K, L, M

N, P, Q

R, S, T

U, W, Z

FSC
www.fsc.org
MIX
Papier | Fördert gute Waldnutzung
FSC® C083411

Kolibri 360 GmbH
Ferdinand-Jühlke-Straße 7
99095 Erfurt, Deutschland
produktsicherheit@kolibri360.de